Praise for Plan B

"Lester Brown tells us how to build a more just world and save the planet... in a practical, straightforward way. We should all heed his advice."

—President Bill Clinton

"... a far-reaching thinker."

—*U.S. News & World Report*

"The best book on the environment I've ever read."

—Chris Swan, *Financial Times*

"It's exciting... a masterpiece!"

—Ted Turner

"[Brown's] ability to make a complicated subject accessible to the general reader is remarkable... "

—Katherine Salant, *Washington Post*

"In tackling a host of pressing issues in a single book, *Plan B 2.0* makes for an eye-opening read."

—*Times Higher Education Supplement*

"A great blueprint for combating climate change."

—Bryan Walsh, *Time*

"[Brown] lays out one of the most comprehensive set of solutions you can find in one place."

—Joseph Romm, *Climate Progress*

"... a highly readable and authoritative account of the problems we face from global warming to shrinking water resources, fisheries, forests, etc. The picture is very frightening. But the book also provides a way forward."

—Clare Short, British Member of Parliament

"Lester R. Brown gives concise, but very informative, summaries of what he regards as the key issues facing civilization as a consequence of the stress we put on our environment.... a valuable contribution to the ongoing debate."

—*The Ecologist*

"Brown is impassioned and convincing when talking about the world's ills and what he considers the four great goals to restoring civilization's equilibrium..."

—April Streeter, *TreeHugger.com*

"In this impressively researched manifesto for change, Brown bluntly sets out the challenges and offers an achievable road map for solving the climate change crisis."

—*The Guardian*

"...the best summation of humanity's converging ecological problems and the best roadmap to solving them, all in one compact package."

—David Roberts, *Grist*

"Lester R. Brown...offers an attractive 21st-century alternative to the unacceptable business-as-usual path that we have been following with regard to the environment (Plan A), which is leading us to 'economic decline and collapse.'"

— Thomas F. Malone, *American Scientist*

"Brown's overall action plan is both comprehensive and compelling."

—Caroline Lucas, *Resurgence*

"Beautifully written and unimpeachably well-informed."

—Ross Gelbspan, author of *The Heat Is On*

"The best single volume on saving the earth, period."

—Geoffrey Holland, author of *The Hydrogen Age*

World on the Edge

OTHER NORTON BOOKS BY LESTER R. BROWN

Plan B 4.0: Mobilizing to Save Civilization

Plan B 3.0: Mobilizing to Save Civilization

Plan B 2.0: Rescuing a Planet Under Stress and a Civilization in Trouble

Outgrowing the Earth: The Food Security Challenge in an Age of Falling Water Tables and Rising Temperatures

Plan B: Rescuing a Planet under Stress and a Civilization in Trouble

The Earth Policy Reader
with Janet Larsen and Bernie Fischlowitz-Roberts

Eco-Economy: Building an Economy for the Earth

State of the World 1984 through *2001*
annual, with others

Vital Signs 1992 through *2001*
annual, with others

Beyond Malthus
with Gary Gardner and Brian Halweil

The World Watch Reader 1998
editor with Ed Ayres

Tough Choices

Who Will Feed China?

Full House
with Hal Kane

Saving the Planet
with Christopher Flavin and Sandra Postel

Building a Sustainable Society

Running on Empty
with Colin Norman and Christopher Flavin

The Twenty-Ninth Day

In the Human Interest

Earth Policy Institute® is a nonprofit environmental research organization providing a plan for building a sustainable future. In addition to the Plan B series, the Institute issues four-page *Plan B Updates* that assess progress in implementing Plan B. All of these plus additional data and graphs can be downloaded at no charge from the EPI Web site.

Web site: www.earth-policy.org

WORLD ON THE EDGE

How to Prevent Environmental and Economic Collapse

Lester R. Brown

EARTH POLICY INSTITUTE

W · W · NORTON & COMPANY

NEW YORK LONDON

Printed in the United States of America
First Edition

The text of this book is composed in Sabon. Composition by Elizabeth Doherty; manufacturing by Courier Westford.

ISBN 978-0-393-08029-2 (cloth) 978-0-393-33949-9 (pbk)

W. W. Norton & Company, Inc., 500 Fifth Avenue,
New York, N.Y. 10110
www.wwnorton.com

W. W. Norton & Company, Ltd., Castle House, 75/76 Wells Street,
London W1T 3QT

1 2 3 4 5 6 7 8 9 0

This book is printed on recycled paper.

Contents

III. THE RESPONSE: PLAN B

IV. WATCHING THE CLOCK

Preface

When I meet old friends and they ask, "How are you?" I often reply, "I'm fine; it's the world I am worried about." "Aren't we all" is the common response. Most people have a rather vague sense of concern about the future, but some worry about specific threats such as climate change or population growth. Some are beyond questioning whether civilization will decline if we continue with business as usual, and instead they are asking when this will occur.

In early 2009, John Beddington, chief science advisor to the U.K. government, said the world was facing a "perfect storm" of food shortages, water scarcity, and costly oil by 2030. These developments, plus accelerating climate change and mass migration across national borders, would lead to major upheavals.

A week later, Jonathon Porritt, former chair of the U.K. Sustainable Development Commission, wrote in the *Guardian* that he agreed with Beddington's analysis but that the timing was off. He thinks the crisis "will hit much closer to 2020 than 2030." He calls it the "ultimate recession"—one from which there may be no recovery.

These assessments by Beddington and Porritt raise two key questions. If we continue with business as usual, how much time do we have left before our global civilization unravels? And how do we save civilization?

World on the Edge is a response to these questions. As to how much time we have left with business as usual, no one knows for sure. We are handicapped by the difficulty of grasping the dynamics of exponential growth in a finite environment—namely, the earth. For me, thinking about this is aided by a riddle the French use to teach schoolchildren exponential growth. A lily pond has one leaf in it the first day, two the second day, four the third, and the number of leaves continues to double each day. If the pond fills on the thirtieth day, when is it half full? The twenty-ninth day. Unfortunately for our overcrowded planet, we may now be beyond the thirtieth day.

My sense is that the "perfect storm" or the "ultimate recession" could come at any time. It will likely be triggered by an unprecedented harvest shortfall, one caused by a combination of crop-withering heat waves and emerging water shortages as aquifers are depleted. Such a grain shortfall could drive food prices off the top of the chart, leading exporting countries to restrict or ban exports—as several countries did when prices rose in 2007–08 and as Russia did again in response to the heat wave of 2010. This in turn would undermine confidence in the market economy as a reliable source of grain. And in a world where each country would be narrowly focused on meeting its own needs, the confidence that is the foundation of the international economic and financial systems would begin to erode.

Now to the second question. What will it take to reverse the many environmental trends that are undermining the world economy? Restructuring the economy in time to avoid decline will take a massive mobilization at wartime speed. Here at the Earth Policy Institute and in this book, we call this massive restructuring Plan B. We are convinced that it, or something very similar to it, is our only hope.

As we think about the ecological deficits that are leading the world toward the edge, it becomes clear that the values generating ecological deficits are the same values that lead to growing fiscal deficits. We used to think it would be our children who would have to deal with the consequences of our deficits, but now it is clear that our generation will have to deal with them. Ecological and economic deficits are now shaping not only our future, but our present.

Beddington and Porritt deserve credit for publicly addressing the prospect of social collapse because it is not easy to talk about. This is partly because it is difficult to imagine something we have never experienced. We lack even the vocabulary. It is also difficult to talk about because we are addressing not just the future of humanity in an abstract sense, but the future of our families and our friends. No generation has faced a challenge with the complexity, scale, and urgency of the one that we face.

But there is hope. Without it this book would not exist. We think we can see both what needs to be done and how to do it.

There are two policy cornerstones underlying the Plan B transformation. One is to restructure taxes by lowering income taxes and raising the tax on carbon emissions to include the indirect costs of burning fossil fuels, such as climate change and air pollution, in fossil fuel prices. The amount of tax we pay would not change.

The second policy cornerstone is to redefine security for the twenty-first century. The threats to our future now are not armed aggression but rather climate change, population growth, water shortages, poverty, rising food prices, and failing states. Our challenge is not only to redefine security in conceptual terms, but also to reallocate fiscal priorities to shift resources toward achieving the Plan B goals. These include reforestation, soil conservation, fishery restoration, universal primary school education, and reproductive health care and family planning services for women everywhere.

Although these goals are conceptually simple and easily understood, they will not be easily achieved. They will require an enormous effort from each of us. The vested interests of the fossil fuel and defense industries in maintaining the status quo are strong. But it is our future that is at stake. Yours and mine.

Lester R. Brown
October 2010

Earth Policy Institute
1350 Connecticut Ave. NW
Suite 403
Washington, DC 20036

Phone: (202) 496-9290
Fax: (202) 496-9325
E-mail: epi@earth-policy.org
Web site: www.earth-policy.org

 For full citations, data, and additional information on the topics discussed in this book, see www.earth-policy.org.

World on the Edge

1

On the Edge

In the summer of 2010, record-high temperatures hit Moscow. At first it was just another heat wave, but the scorching heat that started in late June continued through mid-August. Western Russia was so hot and dry in early August that 300 or 400 new fires were starting every day. Millions of acres of forest burned. So did thousands of homes. Crops withered.

Day after day, Moscow was bathed in seemingly endless smoke. The elderly and those with impaired respiratory systems struggled to breathe. The death rate climbed as heat stress and smoke took their toll.

The average July temperature in Moscow was a scarcely believable 14 degrees Fahrenheit above the norm. Twice during the heat wave, the Moscow temperature exceeded 100 degrees Fahrenheit, a level Muscovites had never before experienced. Watching the heat wave play out over a seven-week period on the TV evening news, with the thousands of fires and the smoke everywhere, was like watching a horror film that had no end. Russia's 140 million people were in shock, traumatized by what was happening to them and their country.

The most intense heat in Russia's 130 years of record-keeping was taking a heavy economic toll. The loss of standing forests and the projected cost of their restoration totaled some $300 billion. Thousands of farmers faced bankruptcy.

Russia's grain harvest shrank from nearly 100 million tons to scarcely 60 million tons as crops withered. Recently the world's number three wheat exporter, Russia banned grain exports in a desperate move to rein in soaring domestic food prices. Between mid-June and mid-August, the world price of wheat climbed 60 percent. Prolonged drought and the worst heat wave in Russian history were boosting food prices worldwide.

But there was some good news coming out of Moscow. On July 30th, Russian President Dmitry Medvedev announced that in large parts of western Russia "practically everything is burning." While sweating, he went on to say, "What's happening with the planet's climate right now needs to be a wake up call to all of us." In something akin to a deathbed conversion, Russia's president was abandoning his country's position as a climate change denier and an opponent of carbon reduction initiatives.

Even before the Russian heat wave ended, there were reports in late July of torrential rains in the mountains of northern Pakistan. The Indus River, the lifeline of Pakistan, and its tributaries were overflowing. Levees that had confined the river to a narrow channel so the fertile floodplains could be farmed had failed. Eventually the raging waters covered one fifth of the country.

The destruction was everywhere. Some 2 million homes were damaged or destroyed. More than 20 million people were affected by the flooding. Nearly 2,000 Pakistanis died. Some 6 million acres of crops were damaged or destroyed. Over a million livestock drowned. Roads and bridges were washed away. Although the flooding was blamed on the heavy rainfall, there were actually several trends converging to produce what was described as the largest natural disaster in Pakistan's history.

On May 26, 2010, the official temperature in Mohenjo-daro in south-central Pakistan reached 128 degrees Fahrenheit, a record for Asia. Snow and glaciers in the western Himalayas, where the tributaries of the Indus River originate, were melting fast. As Pakistani glaciologist M. Iqbal Khan noted, the glacial melt was already swelling the flow of the Indus even before the rains came.

The pressure of population on natural resources is intense. Pakistan's 185 million people are squeezed into an area 8 percent that of the United States. Ninety percent of the original forests in the Indus Basin are gone, leaving little to absorb the rainfall and reduce runoff. Beyond this, Pakistan has a livestock population of cattle, water buffalo, sheep, and goats of 149 million, well above the 103 million grazing livestock in the United States. The result is a country stripped of vegetation. When it rains, rapid runoff erodes the soil, silting up reservoirs and reducing their capacity to store flood water.

Twenty or more years ago, Pakistan chose to define security largely in military terms. When it should have been investing in reforestation, soil conservation, education, and family planning, it was shortchanging these activities to bolster its military capacity. In 1990, the military budget was 15 times that of education and a staggering 44 times that of health and family planning. As a result, Pakistan is now a poor, overpopulated, environmentally devastated nuclear power where 60 percent of women cannot read and write.

What happened to Russia and to Pakistan in the summer of 2010 are examples of what lies ahead for all of us if we continue with business as usual. The media described the heat wave in Russia and the flooding in Pakistan as natural disasters. But were they? Climate scientists have been saying for some time that rising temperatures would bring more extreme climate events. Ecologists have warned that as human pressures on ecosystems mount and as forests and grasslands are destroyed, flooding will be more severe.

The signs that our civilization is in trouble are multiplying. During most of the 6,000 years since civilization began we lived on the sustainable yield of the earth's natural systems. But in recent decades humanity has overshot the level that those systems can sustain.

We are liquidating the earth's natural assets to fuel our consumption. Half of us live in countries where water tables are falling and wells are going dry. Soil erosion exceeds soil formation on one third of the world's cropland, draining the land of its fertility. The world's ever-growing herds of cattle, sheep, and goats are converting vast stretches of grassland to desert. Forests are shrinking by 13 million acres per year as we clear land for agriculture and cut trees for lumber and paper. Four fifths of oceanic fisheries are being fished at capacity or overfished and headed for collapse. In system after system, demand is overshooting supply.

Meanwhile, with our massive burning of fossil fuels, we are overloading the atmosphere with carbon dioxide (CO_2), pushing the earth's temperature ever higher. This in turn generates more frequent and more extreme climatic events, including crop-withering heat waves, more intense droughts, more severe floods, and more destructive storms.

The earth's rising temperature is also melting polar ice sheets and mountain glaciers. If the Greenland ice sheet, which is melting at an accelerating rate, were to melt entirely, it would inundate the rice-growing river deltas of Asia and many of the world's coastal cities. It is the ice melt from the mountain glaciers in the Himalayas and on the Tibetan Plateau that helps sustain the dry-season flow of the major rivers in India and China—the Ganges, Yangtze, and Yellow Rivers—and the irrigation systems that depend on them.

At some point, what had been excessive local demands on environmental systems when the economy was small became global in scope. A 2002 study by a team of scientists led by Mathis Wackernagel aggregates the use of the earth's natural assets, including CO_2 overload in the atmosphere, into a single indicator—the ecological footprint. The authors concluded that humanity's collective demands first surpassed the earth's regenerative capacity around 1980. By 1999, global demands on the earth's natural systems exceeded sustainable yields by 20 percent. Ongoing calculations show it at 50 percent in 2007. Stated otherwise, it would take 1.5 Earths to sustain our current consumption. Environmentally, the world is in overshoot mode. If we use environmental indicators to evaluate our situation, then the global decline of the economy's natural support systems—the environmental decline that will lead to economic decline and social collapse—is well under way.

No previous civilization has survived the ongoing destruction of its natural supports. Nor will ours. Yet economists look at the future through a different lens. Relying heavily on economic data to measure progress, they see the near 10-fold growth in the world economy since 1950 and the associated gains in living standards as the crowning achievement of our modern civilization. During this period, income per person worldwide climbed nearly fourfold, boosting living standards to previously unimaginable levels. A century ago, annual growth in the world economy was measured in the billions of dollars. Today, it is measured in the trillions. In the eyes of mainstream economists, the world has not only an illustrious economic past but also a promising future.

Mainstream economists see the 2008–09 global economic recession and near-collapse of the international financial system as a bump in the road, albeit an unusually big one, before a return to growth as usual. Projections of economic growth, whether by the World Bank, Goldman Sachs, or Deutsche Bank, typically show the global economy expanding by roughly 3 percent a year. At this rate the 2010 economy would easily double in size by 2035. With these projections, economic growth in the decades ahead is more or less an extrapolation of the growth of recent decades.

How did we get into this mess? Our market-based global economy as currently managed is in trouble. The market does many things well. It allocates resources with an efficiency that no central planner could even imagine, much less achieve. But as the world economy expanded some 20-fold over the last century it has revealed a flaw—a flaw so serious that if it is not corrected it will spell the end of civilization as we know it.

The market, which sets prices, is not telling us the truth. It is omitting indirect costs that in some cases now dwarf direct costs. Consider gasoline. Pumping oil, refining it into gasoline, and delivering the gas to U.S. service stations may cost, say, $3 per gallon. The indirect costs, including climate change, treatment of respiratory illnesses, oil spills, and the U.S. military presence in the Middle East to ensure access to the oil, total $12 per gallon. Similar calculations can be done for coal.

We delude ourselves with our accounting system. Leaving such huge costs off the books is a formula for bankruptcy. Environmental trends are the lead indicators telling us what lies ahead for the economy and ultimately for society itself. Falling water tables today signal rising food prices tomorrow. Shrinking polar ice sheets are a prelude to falling coastal real estate values.

Beyond this, mainstream economics pays little attention to the sustainable yield thresholds of the earth's natural systems. Modern economic thinking and policymaking have created an economy that is so out of sync with the ecosystem on which it depends that it is approaching collapse. How can we assume that the growth of an economic system that is shrinking the earth's forests, eroding its soils, depleting its aquifers, collapsing its fisheries, elevating its temperature, and melting its ice sheets can simply be projected into the long-term future? What is the intellectual process underpinning these extrapolations?

We are facing a situation in economics today similar to that in astronomy when Copernicus arrived on the scene, a time when it was believed that the sun revolved around the earth. Just as Copernicus had to formulate a new astronomical worldview after several decades of celestial observations and mathematical calculations, we too must formulate a new economic worldview based on several decades of environmental observations and analyses.

The archeological record indicates that civilizational collapse does not come suddenly out of the blue. Archeologists analyzing earlier civilizations talk about a decline-and-collapse scenario. Economic and social collapse was almost always preceded by a period of environmental decline.

For past civilizations it was sometimes a single environmental trend that was primarily responsible for their decline. Sometimes it was multiple trends. For Sumer, it was rising salt concentrations in the soil as a result of an environmental flaw in the design of their otherwise extraordinary irrigation system. After a point, the salts accumulating in the soil led to a decline in wheat yields. The Sumerians then shifted to barley, a more salt-tolerant crop. But eventually barley yields also began to decline. The collapse of the civilization followed.

Archeologist Robert McC. Adams describes the site of the ancient Sumerian civilization on the central floodplain of the Euphrates River in what is now Iraq as an empty, desolate area now outside the frontiers of cultivation. He says, "Vegetation is sparse, and in many areas it is almost wholly absent....Yet at one time, here lay the core, the heartland, the oldest urban, literate civilization in the world."

For the Mayans, it was deforestation and soil erosion. As more and more land was cleared for farming to support the expanding empire, soil erosion undermined the productivity of their tropical soils. A team of scientists from the National Aeronautics and Space Administration has noted that the extensive land clearing by the Mayans likely also altered the regional climate, reducing rainfall. In effect, the scientists suggest, it was the convergence of several environmental trends, some reinforcing others, that led to the food shortages that brought down the Mayan civilization.

Although we live in a highly urbanized, technologically advanced society, we are as dependent on the earth's natural support systems as the Sumerians and Mayans were. If we continue with business as usual, civilizational collapse is no longer a matter of whether but when. We now have an economy that is destroying its natural support systems, one that has put us on a decline and collapse path. We are dangerously close to the edge. Peter Goldmark, former Rockefeller Foundation president, puts it well: "The death of our civilization is no longer a theory or an academic possibility; it is the road we're on."

Judging by the archeological records of earlier civilizations, more often than not food shortages appear to have precipitated their decline and collapse. Given the advances of modern agriculture, I had long rejected the idea that food could be the weak link in our twenty-first century civilization. Today I think not only that it could be the weak link but that it is the weak link.

The reality of our situation may soon become clearer

for mainstream economists as we begin to see some of the early economic effects of overconsuming the earth's resources, such as rising world food prices. We got a preview when, as world grain demand raced ahead and as supplies tightened in early 2007, the prices of wheat, rice, corn, and soybeans began to climb, tripling historical levels by the spring of 2008. Only the worst global economic downturn since the Great Depression, combined with a record world grain harvest in 2008, managed to check the rise in grain prices, at least for the time being. Since 2008, world market prices have receded somewhat, but as of October 2010, following the disastrous Russian grain harvest, they were still nearly double historical levels and rising.

On the social front, the most disturbing trend is spreading hunger. For the last century's closing decades, the number of chronically hungry and malnourished people worldwide was shrinking, dropping to a low of 788 million by 1996. Then it began to rise—slowly at first, and then more rapidly—as the massive diversion of grain to produce fuel for cars doubled the annual growth in grain consumption. In 2008, it passed 900 million. By 2009, there were more than a billion hungry and malnourished people. The U.N. Food and Agriculture Organization anticipated a decline in the number of hungry people in 2010, but the Russian heat wave and the subsequent climb in grain prices may have ended that hope.

This expansion in the ranks of the hungry is disturbing not only in humanitarian terms but also because spreading hunger preceded collapse for so many of the earlier civilizations whose archeological sites we now study. If we use spreading hunger as an indicator of the decline that precedes social collapse for our global civilization, then it began more than a decade ago.

As environmental degradation and economic and social stresses mount, the more fragile governments are having difficulty managing them. And as rapid population growth continues, cropland becomes scarce, wells go dry, forests disappear, soils erode, unemployment rises, and hunger spreads. In this situation, weaker governments are losing their credibility and their capacity to govern. They become failing states—countries whose governments can no longer provide personal security, food security, or basic social services, such as education and health care. For example, Somalia is now only a place on the map, not a nation state in any meaningful sense of the term.

The term "failing state" has only recently become part of our working vocabulary. Among the many weaker governments breaking down under the mounting stresses are those in Afghanistan, Haiti, Nigeria, Pakistan, and Yemen. As the list of failing states grows longer each year, it raises a disturbing question: How many states must fail before our global civilization begins to unravel?

How much longer can we remain in the decline phase, whether measured in natural asset liquidation, spreading hunger, or failing states, before our global civilization begins to break down? Even as we wrestle with the issues of resource scarcity, world population is continuing to grow. Tonight there will be 219,000 people at the dinner table who were not there last night, many of them with empty plates.

If we continue with business as usual, how much time do we have before we see serious breakdowns in the global economy? The answer is, we do not know, because we have not been here before. But if we stay with business as usual, the time is more likely measured in years than in decades. We are now so close to the edge that it could come at any time. For example, what if the 2010 heat wave centered in Moscow had instead been centered in Chicago? In round numbers, the 40 percent drop from

Russia's recent harvests of nearly 100 million tons cost the world 40 million tons of grain, but a 40-percent drop in the far larger U.S. grain harvest of over 400 million tons would have cost 160 million tons.

While projected world carryover stocks of grain (the amount remaining in the bin when the new harvest begins) for 2011 were reduced from 79 days of world consumption to 72 days by the Russian heat wave, they would have dropped to 52 days of consumption if the heat wave had been centered in Chicago. This level would be not only the lowest on record, but also well below the 62-day carryover that set the stage for the tripling of world grain prices in 2007–08.

In short, if the July temperature in Chicago had averaged 14 degrees above the norm, as it did in Moscow, there would have been chaos in world grain markets. Grain prices would have climbed off the charts. Some grain-exporting countries, trying to hold down domestic food prices, would have restricted or even banned exports, as they did in 2007–08. The TV evening news would be dominated by footage of food riots in low-income grain-importing countries and by reports of governments falling as hunger spread. Grain-importing countries that export oil would be trying to barter oil for grain. Low-income grain importers would lose out. With governments falling and with confidence in the world grain market shattered, the global economy could have started to unravel.

Food price stability now depends on a record or near-record world grain harvest every year. And climate change is not the only threat to food security. Spreading water shortages are also a huge, and perhaps even more imminent, threat to food security and political stability. Water-based "food bubbles" that artificially inflate grain production by depleting aquifers are starting to burst, and as they do, irrigation-based harvests are shrinking. The first food bubble to burst is in Saudi Arabia, where the depletion of its fossil aquifer is virtually eliminating its 3-million-ton wheat harvest. And there are at least another 17 countries with food bubbles based on overpumping.

The Saudi loss of some 3 million tons of wheat is less than 1 percent of the world wheat harvest, but the potential losses in some countries are much larger. The grain produced by overpumping in India feeds 175 million Indians, according to the World Bank. For China, the comparable number is 130 million people. We don't know exactly when these water-based food bubbles will burst, but it could be any time now.

If world irrigation water use has peaked, or is about to, we are entering an era of intense competition for water resources. Expanding world food production fast enough to avoid future price rises will be much more difficult. A global civilization that adds 80 million people each year, even as its irrigation water supply is shrinking, could be in trouble.

When water-based food bubbles burst in larger countries, like China and India, they will push up food prices worldwide, forcing a reduction in consumption among those who can least afford it: those who are already spending most of their income on food. Even now, many families are trying to survive on one meal a day. Those on the lower rungs of the global economic ladder, those even now hanging on by their fingertips, may start to lose their grip.

Further complicating our future, the world may be reaching peak water at more or less the same time that it hits peak oil. Fatih Birol, chief economist with the International Energy Agency, has said, "We should leave oil before it leaves us." I agree. If we can phase out the use of oil quickly enough to stabilize climate, it will also facilitate an orderly, managed transition to a carbon-free renewable energy economy. Otherwise we face intensifying competition among countries for dwindling oil supplies and continued vulnerability to soaring oil prices. And with our recently developed capacity to convert grain into oil (that is, ethanol), the price of grain is now tied to that of oil. Rising oil prices mean rising food prices.

Once the world reaches peak oil and peak water, continuing population growth would mean a rapid drop in the per capita supply of both. And since both are central to food production, the effects on the food supply could leave many countries with potentially unmanageable stresses. And these are in addition to the threats posed by increasing climate volatility. As William Hague, Britain's newly appointed Foreign Secretary and the former leader of the Conservative Party, says, "You cannot have food, water, or energy security without climate security."

Among other things, the situation in which we find ourselves pushes us to redefine security in twenty-first century terms. The time when military forces were the prime threat to security has faded into the past. The threats now are climate volatility, spreading water shortages, continuing population growth, spreading hunger, and failing states. The challenge is to devise new fiscal priorities that match these new security threats.

We are facing issues of near-overwhelming complexity and unprecedented urgency. Can we think systemically and fashion policies accordingly? Can we move fast enough to avoid economic decline and collapse? Can we change direction before we go over the edge?

We are in a race between natural and political tipping points, but we do not know exactly where nature's tipping points are. Nature determines these. Nature is the timekeeper, but we cannot see the clock.

The notion that our civilization is approaching its demise if we continue with business as usual is not an easy concept to grasp or accept. It is difficult to imagine something we have not previously experienced. We hardly have even the vocabulary, much less the experience, to discuss this prospect.

To help us understand how we got so close to the edge, Parts I and II of this book document in detail the trends just described—the ongoing liquidation of the earth's natural assets, the growing number of hungry people, and the lengthening list of failing states.

Since it is the destruction of the economy's natural supports and disruption of the climate system that are driving the world toward the edge, these are the trends that must be reversed. To do so requires extraordinarily demanding measures, a fast shift away from business as usual to what we at the Earth Policy Institute call Plan B. This is described in Part III.

With a scale and urgency similar to the U.S. mobilization for World War II, Plan B has four components: a massive cut in global carbon emissions of 80 percent by 2020; the stabilization of world population at no more than 8 billion by 2040; the eradication of poverty; and the restoration of forests, soils, aquifers, and fisheries.

Carbon emissions can be cut by systematically raising world energy efficiency, by restructuring transport systems, and by shifting from burning fossil fuels to tapping the earth's wealth of wind, solar, and geothermal energy. The transition from fossil fuels to renewable sources of energy can be driven primarily by tax restructuring: steadily lowering income taxes and offsetting this reduction with a rise in the tax on carbon.

Two of the components of Plan B—stabilizing population and eradicating poverty—go hand in hand, reinforcing each other. This involves ensuring at least a primary school education for all children—girls as well as boys. It also means providing at least rudimentary village-level health care so that parents can be more confident that their children will survive to adulthood. And women everywhere need access to reproductive health care and family planning services.

The fourth component, restoring the earth's natural systems and resources, involves, for example, a worldwide initiative to arrest the fall in water tables by raising water productivity. That implies shifting both to more-efficient irrigation systems and to more water-efficient crops. And for industries and cities, it implies doing worldwide what some are already doing—namely, continuously recycling water.

It is time to ban deforestation worldwide, as some countries already have done, and plant billions of trees to sequester carbon. We need a worldwide effort to conserve soil, similar to the U.S. response to the Dust Bowl of the 1930s.

The Earth Policy Institute estimates that stabilizing population, eradicating poverty, and restoring the economy's natural support systems would cost less than $200 billion of additional expenditures a year—a mere one eighth of current world military spending. In effect, the Plan B budget encompassing the measures needed to prevent civilizational collapse is the new security budget.

The situation the world faces now is even more urgent than the economic crisis of 2008 and 2009. Instead of a U.S. housing bubble, it is food bubbles based on overpumping and overplowing that cloud our future. Such food uncertainties are amplified by climate volatility and by more extreme weather events. Our challenge is not just to implement Plan B, but to do it quickly so we can move off the environmental decline path before the clock runs out.

One thing is certain—we are facing greater change than any generation in history. What is not clear is the source of this change. Will we stay with business as usual and enter a period of economic decline and spreading chaos? Or will we quickly reorder priorities, acting at wartime speed to move the world onto an economic path that can sustain civilization?

Data, endnotes, and additional resources can be found on Earth Policy's Web site, at www.earth-policy.org.

I

A DETERIORATING FOUNDATION

2

Falling Water Tables and Shrinking Harvests

The Arab oil-export embargo of the 1970s affected more than just the oil flowing from the Middle East. The Saudis realized that since they were heavily dependent on imported grain, they were vulnerable to a grain counter-embargo. Using oil-drilling technology, they tapped into an aquifer far below the desert to produce irrigated wheat. In a matter of years, Saudi Arabia was self-sufficient in wheat, its principal staple food.

But after more than 20 years of wheat self-sufficiency, the Saudis announced in January 2008 that this aquifer was largely depleted and they would be phasing out wheat production. Between 2007 and 2010, the wheat harvest of nearly 3 million tons dropped by more than two thirds. At this rate the Saudis will harvest their last wheat crop in 2012 and then will be totally dependent on imported grain to feed nearly 30 million people.

The unusually rapid phaseout of wheat farming in Saudi Arabia is due to two factors. First, in this arid country there is little farming without irrigation. Second, irrigation there depends almost entirely on a fossil aquifer, which unlike most aquifers does not recharge naturally from rainfall. The desalted sea water Saudi Arabia uses to supply its cities is far too costly for irrigation use.

Saudi Arabia's growing food insecurity has even led it to buy or lease land in several other countries, including two of the world's hungriest, Ethiopia and Sudan. In effect, the Saudis are planning to produce food for themselves with the land and water resources of other countries.

In neighboring Yemen, replenishable aquifers are being pumped well beyond the rate of recharge, and the deeper fossil aquifers are also being rapidly depleted. As a result, water tables are falling throughout Yemen by some 2 meters per year. In the capital, Sana'a—home to 2 million people—tap water is available only once every 4 days; in Taiz, a smaller city to the south, it is once every 20 days.

Yemen, with one of the world's fastest-growing populations, is becoming a hydrological basket case. With water tables falling, the grain harvest has shrunk by one third over the last 40 years, while demand has continued its steady rise. As a result, the Yemenis now import more than 80 percent of their grain. With its meager oil exports falling, with no industry to speak of, and with nearly 60 percent of its children stunted and chronically undernourished, this poorest of the Arab countries is facing a bleak future.

The likely result of the depletion of Yemen's aquifers—which will lead to further shrinkage of its harvest and spreading hunger and thirst—is social collapse. Already a failing state, it may well devolve into a group of tribal fiefdoms, warring over whatever meager water resources remain. Yemen's internal conflicts could spill over its long, unguarded border with Saudi Arabia.

These two countries represent extreme cases, but many other countries also face dangerous water shortages. The world is incurring a vast water deficit—one that is largely invisible, historically recent, and growing fast. Half the world's people live in countries where water tables are falling as aquifers are being depleted. And since

70 percent of world water use is for irrigation, water shortages can quickly translate into food shortages.

The global water deficit is a product of the tripling of water demand over the last half-century coupled with the worldwide spread of powerful diesel and electrically driven pumps. Only since the advent of these pumps have farmers had the pumping capacity to pull water out of aquifers faster than it is replaced by precipitation.

As the world demand for food has soared, millions of farmers have drilled irrigation wells to expand their harvests. In the absence of government controls, far too many wells have been drilled. As a result, water tables are falling and wells are going dry in some 20 countries, including China, India, and the United States—the three countries that together produce half the world's grain.

The overpumping of aquifers for irrigation temporarily inflates food production, creating a food production bubble, one that bursts when the aquifer is depleted. Since 40 percent of the world grain harvest comes from irrigated land, the potential shrinkage of the supply of irrigation water is of great concern. Among the big three grain producers, roughly a fifth of the U.S. grain harvest comes from irrigated land. For India, the figure is three fifths and for China, roughly four fifths.

There are two sources of irrigation water: underground water and surface water. Most underground water comes from aquifers that are regularly replenished with rainfall; these can be pumped indefinitely as long as water extraction does not exceed recharge. But a distinct minority of aquifers are fossil aquifers—containing water put down eons ago. Since these do not recharge, irrigation ends whenever they are pumped dry. Among the more prominent fossil aquifers are the Ogallala underlying the U.S. Great Plains, the Saudi one described earlier, and the deep aquifer under the North China Plain.

Surface water, in contrast, is typically stored behind dams on rivers and then channeled onto the land through a network of irrigation canals. Historically, and most notably from 1950 until the mid-1970s, when many of the world's large dams were built, this was the main source of growth in world irrigated area. During the 1970s, however, as the sites for new dams became fewer, the growth in irrigated area shifted from building dams to drilling wells in order to gain access to underground water.

Given a choice, farmers prefer having their own wells because they can control the timing and the amount delivered in a way that is not possible with large, centrally managed canal irrigation systems. Pumps let them apply water when the crop needs it, thus achieving higher yields than with large-scale, river-based irrigation systems. As world demand for grain climbed, farmers throughout the world drilled more and more irrigation wells with little concern for how many the local aquifer could support. As a result, water tables are falling and millions of irrigation wells are going dry or are on the verge of doing so.

There are two rather scary dimensions of the emerging worldwide shortage of irrigation water. One is that water tables are falling in many countries at the same time. The other is that once rising water demand climbs above the recharge rate of an aquifer, the excess of demand over sustainable yield widens with each passing year. This means that the drop in the water table as a result of overpumping is also greater each year. Since growth in the demand for water is typically exponential, largely a function of population growth, the decline of the aquifer is also exponential. What starts as a barely noticeable annual drop in the water table can become a rapid fall.

The shrinkage of irrigation water supplies in the big three grain-producing countries is of particular concern.

Thus far, these countries have managed to avoid falling harvests at the national level, but continued overexploitation of aquifers could soon catch up with them. In most of the leading U.S. irrigation states, the irrigated area has peaked and begun to decline. In California, historically the irrigation leader, a combination of aquifer depletion and the diversion of irrigation water to fast-growing cities has reduced irrigated area from nearly 9 million acres in 1997 to an estimated 7.5 million acres in 2010. (One acre equals 0.4 hectares.) In Texas, the irrigated area peaked in 1978 at 7 million acres, falling to some 5 million acres as the Ogallala aquifer underlying much of the Texas panhandle was depleted.

Other states with shrinking irrigated area include Arizona, Colorado, and Florida. Colorado has watched its irrigated area shrink by 15 percent over the last decade or so. Researchers there project a loss of up to 400,000 acres of irrigated land between 2000 and 2030—a drop of more than one tenth. All three states are suffering from both aquifer depletion and the diversion of irrigation water to urban centers. And now that the states that were rapidly expanding their irrigated area, such as Nebraska and Arkansas, are starting to level off, the prospects for any national growth in irrigated area have faded. With water tables falling as aquifers are depleted under the Great Plains and California's Central Valley, and with fast-growing cities in the Southwest taking more and more irrigation water, the U.S. irrigated area has likely peaked.

India is facing a much more difficult situation. A World Bank study reported in 2005 that the grain supply for 175 million Indians was produced by overpumping water. This situation is widespread—with water tables falling and wells going dry in most states. These include Punjab and Haryana, two surplus grain producers that supply most of the wheat and much of the rice used in

India's massive food distribution program for low-income consumers.

Up-to-date and reliable information is not always easy to get. But it is clear that overpumping is extensive, water tables are falling, wells are going dry, and farmers who can afford to are drilling ever deeper wells in what has been described as "a race to the bottom."

Is India's irrigated area still expanding or has it started to shrink? Based on studies by independent researchers, there is little reason to believe that it is still expanding and ample reason to think that in India, as in the United States, decades of overpumping in key states are leading to aquifer depletion on a scale that is reducing the irrigation water supply. India's water-based food bubble may be about to burst.

In China, although surface water is widely used for irrigation, the principal concern is the northern half of the country, where rainfall is low and water tables are falling everywhere. This includes the highly productive North China Plain, which stretches from just north of Shanghai to well north of Beijing and which produces half of the country's wheat and a third of its corn.

Overpumping in the North China Plain suggests that some 130 million Chinese are being fed with grain produced with the unsustainable use of water. Farmers in this region are pumping from two aquifers: the so-called shallow aquifer, which is rechargeable but largely depleted, and the deep fossil aquifer. Once the latter is depleted, the irrigated agriculture dependent on it will end, forcing farmers back to rainfed farming.

A little-noticed groundwater survey done a decade ago by the Geological Environment Monitoring Institute (GEMI) in Beijing reported that under Hebei Province, in the heart of the North China Plain, the average level of the deep aquifer dropped 2.9 meters (nearly 10 feet) in 2000. Around some cities in the province, it fell by 6 meters. He Qingcheng, head of the GEMI groundwater monitoring team, notes that as the deep aquifer under the North China Plain is depleted, the region is losing its last water reserve—its only safety cushion.

In a 2010 interview with *Washington Post* reporter Steven Mufson, He Qingcheng noted that underground water now meets three fourths of Beijing's water needs. The city, he said, is drilling 1,000 feet down to reach water—five times deeper than 20 years ago. His concerns are mirrored in the unusually strong language of a World Bank report on China's water situation that foresees "catastrophic consequences for future generations" unless water use and supply can quickly be brought back into balance.

Furthermore, China's water-short cities and rapidly growing industrial sector are taking an ever-greater share of the available surface and underground water resources. In many situations, growth in urban and industrial demand for water can be satisfied only by diverting water from farmers.

When will China's irrigated area begin to shrink? The answer is not clear yet. Although aquifer depletion and the diversion of water to cities are threatening to reduce the irrigated area in northern China, new dams being built in the mountainous southwest may expand the irrigated area somewhat, offsetting at least some of the losses elsewhere. However, it is also possible that the irrigated area has peaked in China—and therefore in all three of the leading grain-producing countries.

The geographic region where water shortages are most immediately affecting food security is the Middle East. In addition to the bursting food bubble in Saudi Arabia and the fast-deteriorating water situation in Yemen, both Syria and Iraq—the other two populous countries in the region—have water troubles. Some of these arise from the reduced flows of the Euphrates and

Tigris Rivers, which both countries depend on for irrigation water. Turkey, which controls the headwaters of these rivers, is in the midst of a massive dam building program that is slowly reducing downstream flows. Although all three countries are party to water-sharing arrangements, Turkey's ambitious plans to expand both hydropower and irrigation are being fulfilled partly at the expense of its two downstream neighbors.

Given the future uncertainty of river water supplies, farmers in Syria and Iraq are drilling more wells for irrigation. This is leading to overpumping and an emerging water-based food bubble in both countries. Syria's grain harvest has fallen by one fifth since peaking at roughly 7 million tons in 2001. In Iraq, the grain harvest has fallen by one fourth since peaking at 4.5 million tons in 2002.

Jordan, with 6 million people, is also on the ropes agriculturally. Forty or so years ago, it was producing over 300,000 tons of grain annually. Today it produces only 60,000 tons and thus must import over 90 percent of its grain. In this region only Lebanon has avoided a decline in grain production.

In Israel, which banned the irrigation of wheat in 2000 due to water scarcity, production of grain has been falling since 1983. With a population of 7 million people, Israel now imports 98 percent of the grain it consumes.

To the east, water supplies are also tightening in Iran. An estimated one fifth of its 75 million people are being fed with grain produced by overpumping. Iran has the largest food bubble in the region.

Thus in the Middle East, where populations are growing fast, the world is seeing the first collision between population growth and water supply at the regional level. For the first time in history, grain production is dropping in a geographic region with nothing in sight to arrest the decline. Because of the failure of governments in the region to mesh population and water policies, each day now brings 10,000 more people to feed and less irrigation water with which to feed them.

Afghanistan, a country of 29 million people, is also faced with fast-spreading water shortages as water tables fall and wells go dry. In 2008 Sultan Mahmood Mahmoodi, a senior official in the Afghan Ministry of Water and Energy, said "our assessments indicate that due to several factors, mostly drought and excessive use, about 50 percent of groundwater sources have been lost in the past several years." The response is to drill deeper wells, but this only postpones the inevitable day of reckoning—the time when aquifers go dry and the irrigated land reverts to much less productive dryland farming. Drilling deeper treats the symptoms, not the cause, of this issue. Afghanistan, a landlocked country with a fast-growing population, is already importing a third of its grain from abroad.

Thus far the countries where shrinking water resources are measurably reducing grain harvests are all ones with smaller populations. But what about the middle-sized countries such as Pakistan or Mexico, which are also overpumping their aquifers to feed growing populations?

Pakistan, struggling to remain self-sufficient in wheat, appears to be losing the battle. Its population of 185 million in 2010 is projected to reach 246 million by 2025, which means trying to feed 61 million more people in 15 years. But water levels in wells are already falling by a meter or more each year around the twin cities of Islamabad and Rawalpindi. They are also falling under the fertile Punjab plain, which Pakistan shares with India. Pakistan's two large irrigation reservoirs, Mangla and Tarbela, have lost one third of their storage capacity over the last 40 years as they have filled with silt. A World Bank report, *Pakistan's Water Economy: Running Dry*, sums up the situation: "the survival of a modern and growing Pakistan is threatened by water."

In Mexico, home to 111 million people, the demand for water is outstripping supply. Mexico City's water problems are well known, but rural areas are also suffering. In the agricultural state of Guanajuato, the water table is falling by 6 feet or more a year. In the northwestern wheat-growing state of Sonora, farmers once pumped water from the Hermosillo aquifer at a depth of 40 feet. Today, they pump from over 400 feet. With 51 percent of all water extraction in Mexico from aquifers that are being overpumped, Mexico's food bubble may burst soon.

In our water-scarce world, the competition between farmers and cities is intensifying. The economics of water use do not favor farmers in this struggle, simply because it takes so much water to produce food. For example, while it takes only 14 tons of water to produce a ton of steel, it takes 1,000 tons of water to produce a ton of wheat. In countries preoccupied with expanding the economy and creating jobs, agriculture becomes the residual claimant.

Worldwide, roughly 70 percent of all water use is for irrigation, 20 percent goes to industry, and 10 percent goes to residential use. Cities in Asia, the Middle East, and North America are turning to farmers for water. This is strikingly evident in Chennai (formerly Madras), a city of 8 million on the east coast of India. As a result of the city government's inability to supply water to many of its residents, a thriving tank-truck industry has emerged that buys water from farmers and hauls it to the city's thirsty residents.

For farmers near the city, the market price of water far exceeds the value of the crops they can produce with it. Unfortunately, the 13,000 tankers hauling water to Chennai are mining the region's underground water resources. Water tables are falling and shallow wells have gone dry. Eventually even the deeper wells will go dry, depriving these communities of both their food supply and their livelihood.

In the U.S. southern Great Plains and the Southwest, where water supplies are tight, the growing water needs of cities and thousands of small towns can be satisfied only by taking water from agriculture. A monthly publication from California, the *Water Strategist*, devotes several pages to a listing of water sales in the western United States. Scarcely a working day goes by without another sale. A University of Arizona study of over 2,000 of these water transfers from 1987 to 2005 reported that at least 8 out of 10 involved individual farmers or irrigation districts selling water to cities and municipalities.

Colorado has one of the world's most active water markets. Fast-growing cities and towns in a state with high immigration are buying irrigation water rights from farmers and ranchers. In the upper Arkansas River basin, which occupies the southeastern quarter of the state, Colorado Springs and Aurora (a suburb of Denver) have already bought water rights to one third of the basin's farmland. Aurora has purchased rights to water that was once used to irrigate 23,000 acres of cropland in Colorado's Arkansas Valley.

Even larger purchases are being made by cities in California. In 2003, San Diego bought annual rights to 247 million tons (200,000 acre-feet) of water from farmers in the nearby Imperial Valley—the largest farm-to-city water transfer in U.S. history. This agreement covers the next 75 years. And in 2004, the Metropolitan Water District, which supplies water to some 19 million southern Californians in several cities, negotiated the purchase of 137 million tons of water per year from farmers for the next 35 years. Without irrigation water, the highly productive land owned by these farmers is wasteland. The sellers would like to continue farming, but city officials offer far more for the water than the farmers could possibly earn by irrigating crops.

Whether it is outright government expropriation, farmers being outbid by cities, or cities simply drilling deeper wells than farmers can afford, the world's farmers are losing the water war. For them, it is all too often a shrinking share of a shrinking supply. Slowly but surely, fast-growing cities are siphoning water from the world's farmers even as they try to feed some 80 million more people each year.

In countries where virtually all water is spoken for, as in North Africa and the Middle East, cities can typically get more water only by taking it from irrigation. Countries then import grain to offset the loss of grain production. Since it takes 1,000 tons of water to produce 1 ton of grain, importing grain is the most efficient way to import water. Countries are in effect using grain to balance their water books. Similarly, trading in grain futures is, in a sense, trading in water futures. To the extent that there is a world water market, it is embodied in the world grain market.

How are all these pressures on water supplies affecting grain production in individual countries and worldwide? Is irrigated area expanding or shrinking? If the latter, is it shrinking fast enough to override technological gains and reduce the grain harvest in absolute terms, or will it simply slow its growth?

Today more than half of the world's people live in countries with food bubbles. The question for each of these countries is not whether its bubble will burst, but when—and how the government will cope with it. Will governments be able to import grain to offset production losses? For some countries, the bursting of the bubble may well be catastrophic. For the world as a whole, the near-simultaneous bursting of several national food bubbles as aquifers are depleted could create unmanageable food shortages.

This situation poses an imminent threat to food security and political stability. We have a choice to make. We can continue with overpumping as usual and suffer the consequences. Or we can launch a worldwide effort to stabilize aquifers by raising water productivity—patterning the campaign on the highly successful effort to raise grainland productivity that was launched a half-century ago.

Data, endnotes, and additional resources can be found on Earth Policy's Web site, at www.earth-policy.org.

3

Eroding Soils and Expanding Deserts

On March 20th, 2010, a suffocating dust storm enveloped Beijing. The city's weather bureau took the unusual step of describing the air quality as hazardous, urging people to stay inside or to cover their faces when they were outdoors. Visibility was low, forcing motorists to drive with their lights on in daytime.

Beijing was not the only area affected. This particular dust storm engulfed scores of cities in five provinces, directly affecting over 250 million people. It was not an isolated incident. Every spring, residents of eastern Chinese cities, including Beijing and Tianjin, hunker down as the dust storms begin. Along with the difficulty in breathing and the dust that stings the eyes, there is a constant struggle to keep dust out of homes and to clear doorways and sidewalks of dust and sand. The farmers and herders whose livelihoods are blowing away are paying an even higher price.

These annual dust storms affect not only China, but neighboring countries as well. The March 20th dust storm arrived in South Korea soon after leaving Beijing. It was described by the Korean Meteorological Administration (KMA) as the worst dust storm on record.

In a detailed account in the *New York Times*, Howard French described a Chinese dust storm that reached

Korea on April 12, 2002. South Korea, he said, was engulfed by so much dust from China that people in Seoul were literally gasping for breath. Schools were closed, airline flights were cancelled, and clinics were overrun with patients having difficulty breathing. Retail sales fell. Koreans have come to dread the arrival of what they call "the fifth season"—the dust storms of late winter and early spring.

And the situation continues to deteriorate. The KMA reports that Seoul has "suffered 'dust events' on 23 days during the 1970s, 41 days in the 1980s, 70 days in the 1990s, and 96 days so far this decade."

While people living in China and South Korea are all too familiar with dust storms, the rest of the world typically learns about this fast-growing ecological catastrophe when the massive soil-laden storms leave the region. On April 18, 2001, for instance, the western United States—from the Arizona border north to Canada—was blanketed with dust. It came from a huge dust storm that originated in northwestern China and Mongolia on April 5th.

Nine years later, in April 2010, a National Aeronautics and Space Administration (NASA) satellite tracked a dust storm from China as it journeyed to the east coast of the United States. Originating in the Taklimakan and Gobi Deserts, it ultimately covered an area stretching from North Carolina to Pennsylvania. Each of these huge dust storms carried millions of tons of China's topsoil, a resource that will take centuries to replace.

The thin layer of topsoil that covers much of the earth's land surface and is typically measured in inches is the foundation of civilization. Geomorphologist David Montgomery, in *Dirt: The Erosion of Civilizations*, describes soil as "the skin of the earth—the frontier between geology and biology." After the earth was created, soil formed slowly over geological time from the weathering of rocks. It was this soil that supported early plant life on land. As plant life spread, the plants protected the soil from wind and water erosion, permitting it to accumulate and to support even more vegetation. This relationship facilitated an accumulation of topsoil that could support a rich diversity of plant and animal life.

As long as soil erosion on cropland does not exceed new soil formation, all is well. But once it does, it leads to falling soil fertility and eventually to land abandonment. Sadly, soil formed on a geological time scale is being removed on a human time scale.

Journalist Stephen Leahy writes in *Earth Island Journal* that soil erosion is "the silent global crisis." He notes that "it is akin to tire wear on your car—a gradual, unobserved process that has potentially catastrophic consequences if ignored for too long."

Losing productive topsoil means losing both organic matter in the soil and vegetation on the land, thus releasing carbon into the atmosphere. Rattan Lal, a soil scientist at Ohio State University, notes that the 2,500 billion tons of carbon stored in soils dwarfs the 760 billion tons in the atmosphere. The bottom line is that land degradation is helping drive climate change.

Soil erosion is not new. It is as old as the earth itself. What is new is that it has gradually accelerated ever since agriculture began. At some point, probably during the nineteenth century, the loss of topsoil from erosion surpassed the new soil that is formed through natural processes.

Today, roughly a third of the world's cropland is losing topsoil at an excessive rate, thereby reducing the land's inherent productivity. An analysis of several studies on soil erosion's effect on U.S. crop yields concluded that for each inch of topsoil lost, wheat and corn yields declined by close to 6 percent.

In August 2010, the United Nations announced that

desertification now affects 25 percent of the earth's land area. And it threatens the livelihoods of more than 1 billion people—the families of farmers and herders in roughly 100 countries.

Dust storms provide highly visible evidence of soil erosion and desertification. Once vegetation is removed either by overgrazing or overplowing, the wind begins to blow the small soil particles away. Because the particles are small, they can remain airborne over great distances. Once they are largely gone, leaving only larger particles, sand storms begin. These are local phenomena, often resulting in dune formation and the abandonment of both farming and grazing. Sand storms are the final phase in the desertification process.

In some situations, the threat to topsoil comes primarily from overplowing, as in the U.S. Dust Bowl, but in other situations, such as in northern China, the cause is primarily overgrazing. In either case, permanent vegetation is destroyed and soils become vulnerable to both wind and water erosion.

Giant dust bowls are historically new, confined to the last century or so. During the late nineteenth century, millions of Americans pushed westward, homesteading on the Great Plains, plowing vast areas of grassland to produce wheat. Much of this land—highly erodible when plowed—should have remained in grass. Exacerbated by a prolonged drought, this overexpansion culminated in the 1930s Dust Bowl, a traumatic period chronicled in John Steinbeck's novel *The Grapes of Wrath*. In a crash program to save its soils, the United States returned large areas of eroded cropland to grass, adopted strip-cropping, and planted thousands of miles of tree shelterbelts.

Three decades later, history repeated itself in the Soviet Union. In an all-out effort to expand grain production in the late 1950s, the Soviets plowed an area of grassland roughly equal to the wheat area of Australia and Canada combined. The result, as Soviet agronomists had predicted, was an ecological disaster—another Dust Bowl.

Kazakhstan, which was at the center of this Soviet Virgin Lands Project, saw its grainland area peak at just over 25 million hectares in the mid-1980s. (One hectare equals 2.47 acres.) It then shrank to less than 11 million hectares in 1999. It is now slowly expanding, and grainland area is back up to 17 million hectares. Even on the remaining land, however, the average wheat yield is scarcely 1 ton per hectare, a far cry from the 7 tons per hectare that farmers get in France, Western Europe's leading wheat producer.

Today, two giant dust bowls are forming. One is in the Asian heartland in northern and western China, western Mongolia, and central Asia. The other is in central Africa in the Sahel—the savannah-like ecosystem that stretches across Africa, separating the Sahara Desert from the tropical rainforests to the south. Both are massive in scale, dwarfing anything the world has seen before. They are caused, in varying degrees, by overgrazing, overplowing, and deforestation.

China may face the biggest challenge of all. After the economic reforms in 1978 that shifted the responsibility for farming from large state-organized production teams to individual farm families, China's cattle, sheep, and goat populations spiraled upward. The United States, a country with comparable grazing capacity, has 94 million cattle, a slightly larger herd than China's 92 million. But when it comes to sheep and goats, the United States has a combined population of only 9 million, whereas China has 281 million. Concentrated in China's western and northern provinces, these animals are stripping the land of its protective vegetation. The wind then does the rest, removing the soil and converting rangeland into desert.

Wang Tao, one of the world's leading desert scholars, reports that from 1950 to 1975 an average of 600 square miles of land turned to desert each year. Between 1975 and 1987, this climbed to 810 square miles a year. From then until the century's end, it jumped to 1,390 square miles of land going to desert annually.

China is now at war. It is not invading armies that are claiming its territory, but expanding deserts. Old deserts are advancing and new ones are forming like guerrilla forces striking unexpectedly, forcing Beijing to fight on several fronts.

A U.S. Embassy report entitled "Desert Mergers and Acquisitions" describes satellite images showing two deserts in north-central China expanding and merging to form a single, larger desert overlapping Inner Mongolia and Gansu Provinces. To the west in Xinjiang Province, two even larger deserts—the Taklimakan and Kumtag—are also heading for a merger. Highways running through the shrinking region between them are regularly inundated by sand dunes.

While major dust storms make the news when they affect cities, the heavy damage is in the area of origin. These regions are affected by storms of dust and sand combined. A scientific paper describes in vivid detail a 1993 sandstorm in Gansu Province in China's northwest. This intense sand and dust storm reduced visibility to zero, and the daytime sky was described as "dark as a winter night." It destroyed 430,000 acres of standing crops, damaged 40,000 trees, killed 67,000 cattle and sheep, blew away 67,000 acres of plastic greenhouses, injured 278 people, and killed 49 individuals. Forty-two passenger and freight trains were either cancelled, delayed, or simply parked to wait until the storm passed and the tracks were cleared of sand dunes.

While China is battling its expanding deserts, India, with scarcely 2 percent of the world's land area, is struggling to support 17 percent of the world's people and 18 percent of its cattle. According to a team of scientists at the Indian Space Research Organization, 24 percent of India's land area is slowly turning into desert. It thus comes as no surprise that many of India's cattle are emaciated and over 40 percent of its children are chronically hungry and underweight.

Africa, too, is suffering heavily from unsustainable demands on its croplands and grasslands. Rattan Lal made the first estimate of continental yield losses due to soil erosion. He concluded that soil erosion and other forms of land degradation have cost Africa 8 million tons of grain per year, or roughly 8 percent of its annual harvest. Lal expects the loss to climb to 16 million tons by 2020 if soil erosion continues unabated.

On the northern fringe of the Sahara, countries such as Algeria and Morocco are attempting to halt the desertification that is threatening their fertile croplands. Algerian president Abdelaziz Bouteflika says that Algeria is losing 100,000 acres of its most fertile lands to desertification each year. For a country that has only 7 million acres of grainland, this is not a trivial loss. Among other measures, Algeria is planting its southernmost cropland in perennials, such as fruit orchards, olive orchards, and vineyards—crops that can help keep the soil in place.

Mounting population pressures are evident everywhere on this continent where the growth in livestock numbers closely tracks that in human numbers. In 1950, Africa was home to 227 million people and about 300 million livestock. By 2009, there were 1 billion people and 862 million livestock. With livestock demands now often exceeding grassland carrying capacity by half or more, grassland is turning into desert. In addition to overgrazing, parts of the Sahel are suffering from an extended drought, one that scientists link to climate change.

There is no need to visit soil-devastated countries in order to see the evidence of severe erosion in Africa. Dust storms originating in the new dust bowls are now faithfully recorded in satellite images. On January 9, 2005, NASA released images of a vast dust storm moving westward out of central Africa. This huge cloud of tan-colored dust extended over 3,300 miles—enough to stretch across the United States from coast to coast.

Andrew Goudie, professor of geography at Oxford University, reports that the incidence of Saharan dust storms—once rare—has increased 10-fold during the last half-century. Among the African countries most affected by soil loss from wind erosion are Niger, Chad, Mauritania, northern Nigeria, and Burkina Faso. In Mauritania, in Africa's far west, the number of dust storms jumped from 2 a year in the early 1960s to 80 a year recently.

The Bodélé Depression in Chad is the source of an estimated 1.3 billion tons of wind-borne soil a year, up 10-fold since measurements began in 1947. The nearly 3 billion tons of fine soil particles that leave Africa each year in dust storms are slowly draining the continent of its fertility and biological productivity. In addition, dust storms leaving Africa travel westward across the Atlantic, depositing so much dust in the Caribbean that they cloud the water and damage coral reefs.

Nigeria, Africa's most populous country, reports losing 867,000 acres of rangeland and cropland to desertification each year. While Nigeria's human population was growing from 37 million in 1950 to 151 million in 2008, a fourfold expansion, its livestock population grew from 6 million to 104 million, a 17-fold jump. With the forage needs of Nigeria's 16 million cattle and 88 million sheep and goats exceeding the sustainable yield of grasslands, the northern part of the country is slowly turning to desert. If Nigeria's population keeps growing as projected, the associated land degradation will eventually undermine herding and farming.

In East Africa, Kenya is being squeezed by spreading deserts. Desertification affects up to a fourth of the country's 39 million people. As elsewhere, the combination of overgrazing, overcutting, and overplowing is eroding soils, costing the country valuable productive land.

In Afghanistan, a U.N. Environment Programme (UNEP) team reports that in the Sistan region "up to 100 villages have been submerged by windblown dust and sand." The Registan Desert is migrating westward, encroaching on agricultural areas. In the country's northwest, sand dunes are moving onto agricultural land in the upper Amu Darya basin, their path cleared by the loss of stabilizing vegetation due to firewood gathering and overgrazing. The UNEP team observed sand dunes as high as a five-story building blocking roads, forcing residents to establish new routes.

An Afghan Ministry of Agriculture and Food report reads like an epitaph on a gravestone: "Soil fertility is declining,...water tables have dramatically fallen, de-vegetation is extensive and soil erosion by water and wind is widespread." After nearly three decades of armed conflict and the related deprivation and devastation, Afghanistan's forests are nearly gone. Seven southern provinces are losing cropland to encroaching sand dunes. And like many failing states, even if Afghanistan had appropriate environmental policies, it lacks the law enforcement authority to implement them.

Neighboring Iran illustrates the pressures facing the Middle East. With 8 million cattle and 79 million sheep and goats—the source of wool for its fabled Persian carpet-making industry—Iran's rangelands are deteriorating from overstocking. In the southeastern province of Sistan-Balochistan, sand storms have buried 124 villages, forcing their abandonment. Drifting sands have covered grazing areas, starving livestock and depriving villagers of their livelihood.

In Iraq, suffering from nearly a decade of war and recent drought, a new dust bowl appears to be forming.

Chronically plagued by overgrazing and overplowing, Iraq is now losing irrigation water to its upstream riparian neighbors—Turkey, Syria, and Iran. The reduced river flow—combined with the drying up of marshlands, the deterioration of irrigation infrastructure, and the shrinking irrigated area—is drying out Iraq. The Fertile Crescent, the cradle of civilization, may be turning into a dust bowl.

Dust storms are occurring with increasing frequency in Iraq. In July 2009 a dust storm raged for several days in what was described as the worst such storm in Iraq's history. As it traveled eastward into Iran, the authorities in Tehran closed government offices, private offices, schools, and factories. Although this new dust bowl is small compared with those centered in northwest China and central Africa, it is nonetheless an unsettling new development in this region.

One indicator that helps us assess grassland health is changes in the goat population relative to those of sheep and cattle. As grasslands deteriorate, grass is typically replaced by desert shrubs. In such a degraded environment, cattle and sheep do not fare well, but goats—being particularly hardy ruminants—forage on the shrubs. Between 1970 and 2009, the world cattle population increased by 28 percent and the sheep population stayed relatively static, but the goat population more than doubled.

In some developing countries, the growth in the goat population is dramatic. While Pakistan's cattle population doubled between 1961 and 2009, and the sheep population nearly tripled, the goat population grew more than sixfold and is now equal to that of the cattle and sheep populations combined.

As countries lose their topsoil, they eventually lose the capacity to feed themselves. Among those facing this problem are Lesotho, Haiti, Mongolia, and North Korea.

Lesotho, one of Africa's smallest countries, with only

2 million people, is paying a heavy price for its soil losses. A U.N. team visited in 2002 to assess its food prospect. Their finding was straightforward: "Agriculture in Lesotho faces a catastrophic future; crop production is declining and could cease altogether over large tracts of country if steps are not taken to reverse soil erosion, degradation, and the decline in soil fertility."

Michael Grunwald reported in the *Washington Post* that nearly half of the children under five in Lesotho are stunted physically. "Many," he wrote, "are too weak to walk to school." During the last 10 years, Lesotho's grain harvest dropped by half as its soil fertility fell. Its collapsing agriculture has left the country heavily dependent on food imports.

In the western hemisphere, Haiti—one of the early failing states—was largely self-sufficient in grain 40 years ago. Since then it has lost nearly all its forests and much of its topsoil, forcing it to import over half of its grain. Lesotho and Haiti are both dependent on U.N. World Food Programme lifelines.

A similar situation exists in Mongolia, where over the last 20 years nearly three fourths of the wheatland has been abandoned and wheat yields have started to fall, shrinking the harvest by four fifths. Mongolia now imports nearly 70 percent of its wheat.

North Korea, largely deforested and suffering from flood-induced soil erosion and land degradation, has watched its yearly grain harvest fall from a peak of 5 million tons during the 1980s to scarcely 3.5 million tons during the first decade of this century.

Soil erosion is taking a human toll. Whether the degraded land is in Haiti, Lesotho, Mongolia, North Korea, or any of the many other countries losing their soil, the health of the people cannot be separated from the health of the land itself.

4

Rising Temperatures, Melting Ice, and Food Security

On August 5th, 2010, the Petermann Glacier on the northwest coast of Greenland gave birth to an iceberg that covered 97 square miles. Four times the size of Manhattan, in late 2010 this "ice island" is floating between Greenland and Canada, drifting slowly southward with the prevailing currents. Since it is up to half the height of the Empire State Building in thickness, it could take years for it to melt, break up, and eventually disappear.

News of this massive ice break focused attention on the Greenland ice sheet once more. Scientists have been reporting for some years that it was melting at an accelerating rate. In 2007, Robert Corell, chairman of the Arctic Climate Impact Assessment, reported from Greenland that "we have seen a massive acceleration of the speed with which these glaciers are moving into the sea." He noted that ice was moving at over 6 feet an hour on a front 3 miles long and 1 mile deep.

In August 2010, Richard Bates, a member of a British-led expedition monitoring the Greenland ice sheet, said, "This year marks yet another record-breaking melt year in Greenland; temperatures and melt across the entire ice sheet have exceeded those... of historical records."

Greenland was not alone in experiencing extremes in 2010. New high-temperature records were set in 18 countries. The number of record highs was itself a record, topping the previous total of 15 set in 2007. When a site in south central Pakistan hit 128 degrees Fahrenheit on May 26th, it set not only a new national record, but also a new all-time high for Asia.

Within the United States, numerous cities on the East Coast suffered through the hottest June to August on record, including New York, Philadelphia, and Washington. After a relatively cool summer in Los Angeles, the temperature there on September 27th reached an all-time high of 113 degrees before the official thermometer broke. At a nearby site, however, the thermometer survived to register 119 degrees, a record for the region. What U.S. climate data show us is that as the earth has warmed, record highs are now twice as likely as record lows.

The pattern of more-intense heat waves, more-powerful storms, and more-destructive flooding is consistent with what climate models project will happen as the earth's temperature rises. The worst heat wave in Russian history and the worst flooding in Pakistan's history are the kind of extreme events we can expect to see more of if we continue with business as usual. James Hansen, the U.S. government's leading climate scientist, asks, "Would these events have occurred if atmospheric carbon dioxide had remained at its pre-industrial level of 280 ppm [parts per million]?" The answer, he says, is "almost certainly not."

As atmospheric carbon dioxide levels rise, we can expect even higher temperatures in the future. The earth's average temperature has risen in each of the last four decades, with the increase in the last decade being the largest. As a general matter, temperature rise is projected to be greater in the higher latitudes than in equatorial regions, greater over land than over the oceans, and greater in the interior of continents than in coastal regions.

As the planet heats up, climate patterns shift. Overall, higher temperature means more evaporation and therefore more precipitation. Some parts of the earth will get wetter, other parts dryer. Monsoon patterns will change. Wetter regions will be concentrated in the higher latitudes—including Canada, northern Europe, and Russia—and in Southeast Asia. Places at particular risk of drying out include the Mediterranean region, Australia, and the U.S. Southwest.

Climate instability is becoming the new norm. The time when we could use climate trends of the recent past as a guide to future climate conditions is now history. We are moving into an age of unpredictability.

The effects of high temperatures on food security are scary. Agriculture as it exists today has evolved over 11,000 years of rather remarkable climate stability. As a result, world agriculture has evolved to maximize productivity within this climatic regime. With the earth's climate changing, agriculture will increasingly be out of sync with the climate system that shaped it.

When temperatures soar during the growing season, grain yields fall. Crop ecologists use a rule of thumb that for each 1-degree-Celsius rise in temperature above the optimum during the growing season, we can expect a 10-percent decline in grain yields.

Among other things, temperature affects photosynthesis. In a study of local ecosystem sustainability, Mohan Wali and his colleagues at Ohio State University noted that as temperature rises, photosynthetic activity in plants increases until the temperature reaches 68 degrees Fahrenheit. The rate of photosynthesis then plateaus until the temperature hits 95 degrees, whereupon it begins to decline. At 104 degrees, photosynthesis ceases entirely.

The most vulnerable part of a plant's life cycle is the pollination period. Of the world's three food staples—

rice, wheat, and corn—corn is particularly vulnerable to heat stress. In order for corn to reproduce, pollen must fall from the tassel to the strands of silk that emerge from the end of each ear of corn. Each of these silk strands is attached to a kernel site on the cob. If the kernel is to develop, a grain of pollen must fall on the silk strand and then journey to the kernel site. When temperatures are uncommonly high, the silk strands quickly dry out and turn brown, unable to play their role in the fertilization process.

The effects of temperature on rice pollination have been studied in detail in the Philippines. Scientists there report that the pollination of rice falls from 100 percent at 93 degrees Fahrenheit to nearly zero at 104 degrees Fahrenheit, leading to crop failure.

Heat waves clearly can decimate harvests. Other effects of higher temperatures on our food supply are less obvious but no less serious. Rising temperatures are already melting ice caps and glaciers around the globe. The massive West Antarctic and Greenland ice sheets are both melting. The Greenland ice cap is melting so fast in places that it is triggering minor earthquakes as huge chunks of ice weighing millions of tons break off and slide into the sea.

The breakup of ice in West Antarctica is also gaining momentum. One of the first signals that this ice sheet was breaking up came in 1995 when Larsen A—a huge shelf on the Antarctic Peninsula—collapsed. Then in March 2002 the Larsen B ice shelf collapsed into the sea. At about the same time, over 2,000 square miles of ice broke off the Thwaites Glacier. And in January 2010 an area larger than Rhode Island broke off the nearby Ronne-Filchner ice shelf. If the West Antarctic ice sheet were to melt entirely, sea level would rise by 16 feet.

Temperatures are rising much faster in the Arctic than elsewhere. Winter temperatures in the Arctic, including

Alaska, western Canada, and eastern Russia, have climbed by 4–7 degrees Fahrenheit over the last half-century. This record rise in temperature in the Arctic region could lead to changes in climate patterns that will affect the entire planet.

Sea ice in the Arctic Ocean has been shrinking for the last few decades. Some scientists now think the Arctic Ocean could be free of ice during the summer by 2015—less than five years from now. This worries climate scientists because of the albedo effect. When incoming sunlight strikes the ice in the Arctic Ocean, up to 70 percent is reflected back into space and as little as 30 percent is absorbed as heat. As the Arctic sea ice melts, however, and the incoming sunlight hits the much darker open water, only 6 percent is reflected back into space and 94 percent is converted into heat. This creates a positive feedback—a situation where a trend, once under way, feeds on itself.

If ice disappears entirely in summer and is reduced in winter, the Arctic region will heat up even more, ensuring that the Greenland ice sheet will melt even faster. Recent studies indicate that a combination of melting ice sheets and glaciers, plus the thermal expansion of the ocean as it warms, could raise sea level by up to 6 feet during this century, up from a 6-inch rise during the last century.

Even a 3-foot rise in sea level would sharply reduce the rice harvest in Asia, home to over half of the world's people. It would inundate half the riceland in Bangladesh, a country of 164 million people, and would submerge part of the Mekong Delta, a region that produces half of Viet Nam's rice. Viet Nam, second only to Thailand as a rice exporter, could lose its exportable surplus of rice. This would leave the 20 or so countries that import rice from Viet Nam looking elsewhere.

In addition to the Gangetic Delta in Bangladesh and the Mekong Delta in Viet Nam, numerous other rice-growing river deltas in Asia would be submerged in varying degrees by a 3-foot rise in sea level. It is not intuitively obvious that ice melting on a large island in the far North Atlantic could shrink the rice harvest in Asia, a region that grows 90 percent of the world's rice.

While the ice sheets are melting, so too are mountain glaciers—nature's freshwater reservoirs. The snow and ice masses in the world's mountain ranges and the water they store are taken for granted simply because they have been there since before agriculture began. Now that is changing. If we continue raising the earth's temperature, we risk losing the "reservoirs in the sky" on which so many farmers and cities depend.

Americans need not go far from home to see massive glacier melting. In 1910, when Glacier National Park in western Montana was created, it had some 150 glaciers. In recent decades, these glaciers have been disappearing. By the end of 2009, only 27 were left. In April 2010 park officials announced that 2 more had melted, leaving only 25. It appears to be only a matter of time until all the park's glaciers are gone.

Other landmarks, such as the glaciers on Mount Kilimanjaro in East Africa, are also melting quickly. Between 1912 and 2007, Kilimanjaro's glaciers shrank 85 percent. It is too late to save this landmark. Like the glaciers in Glacier National Park, those on Kilimanjaro may soon be relegated to photographs in museums.

The World Glacier Monitoring Service has reported the nineteenth consecutive year of shrinking mountain glaciers. Glaciers are melting in all of the world's major mountain ranges, including the Andes, the Rockies, the Alps, the Himalayas, and the Tibetan Plateau.

Ice melt from mountain glaciers in the Himalayas and on the Tibetan Plateau helps sustain the major rivers of Asia during the dry season, when irrigation water needs are greatest. In the Indus, Ganges, Yellow, and Yangtze

River basins, where irrigated agriculture depends heavily on the rivers, the loss of any dry-season flow is bad news for farmers.

These melting glaciers coupled with the depletion of aquifers present the most massive threat to food security the world has ever faced. China is the world's leading producer of wheat. India is number two. (The United States is number three.) With rice, China and India totally dominate the world harvest.

In India, the giant Gangotri Glacier, which helps keep the Ganges River flowing during the dry season, is retreating. The Ganges River is not only by far the largest source of surface water irrigation in India, it is also a source of water for the 407 million people living in the Gangetic basin.

Yao Tandong, a leading Chinese glaciologist, reports that glaciers on the Tibetan Plateau in western China are now melting at an accelerating rate. Many smaller glaciers have already disappeared. Yao believes that two thirds of these glaciers could be gone by 2060. If this melting of glaciers continues, Yao says it "will eventually lead to an ecological catastrophe."

The Yellow River basin is home to 147 million people; their fate is closely tied to the river because of low rainfall in the northern half of China. The Yangtze is by far the country's largest river, helping to produce half or more of its 130-million-ton rice harvest. The Yangtze basin is home to 369 million people—more than the entire population of the United States.

Thus the number of people affected by the melting and eventual disappearance of glaciers will be huge. The prospect of shrinking dry-season river flows is unfolding against a startling demographic backdrop: by 2030, India is projected to add 270 million people to its population of 1.2 billion and China is due to add 108 million to its 1.3 billion. While farmers in China and India are already losing irrigation water as overpumping depletes aquifers, they are also facing a reduction of river water for irrigation.

In a world where grain prices have recently climbed to record highs, any disruption of the wheat or rice harvests due to water shortages in India or China will raise their grain imports, driving up food prices. In each of these countries, food prices will likely rise as glaciers disappear and dry-season flows diminish. In India, where just over 40 percent of all children under five years of age are underweight and undernourished, hunger will intensify and child mortality will likely climb.

The depletion of glaciers in the early stage can expand river flows for a time, thus potentially increasing the water available for irrigation. Like the depletion of aquifers, the melting of glaciers can artificially inflate food production for a short period. At some point, however, as the glaciers shrink and the smaller ones disappear entirely, so does the water available for irrigation.

In South America, some 22 percent of Peru's glacial endowment, which feeds the many rivers that supply water to farmers and cities in the arid coastal regions, has disappeared. Ohio State University glaciologist Lonnie Thompson reported in 2007 that the Quelccaya Glacier in southern Peru, which had been retreating by 20 feet per year in the 1960s, was retreating by 200 feet annually. In an interview with *Science News* in early 2009 he said, "It is now retreating up the mountainside by about 18 inches a day, which means you can almost sit there and watch it lose ground."

As Peru's glaciers shrink, the water flow from the mountains to the country's arid coastal region, where 60 percent of the people live, will decline during the dry season. This region includes Lima, which, with nearly 9 million inhabitants, is the world's second largest desert city, after Cairo. Given the coming decline in its water supply, a U.N. study refers to Lima as "a crisis waiting to happen."

Bolivia is also fast losing the glaciers whose ice melt supplies its farmers and cities with water. Between 1975 and 2006, the area of its glaciers shrank by nearly half. Bolivia's famed Chacaltaya Glacier, once the site of the world's highest ski resort, disappeared in 2009.

For the 53 million people living in Peru, Bolivia, and Ecuador, the loss of their mountain glaciers and dry-season river flow threatens their food security and political stability. Not only do farmers in the region produce much of their wheat and potatoes with the river water from these disappearing glaciers, but well over half the region's electricity supply comes from hydroelectric sources. Currently, few countries are being affected by melting mountain glaciers as much as these Andean societies.

In many of the world's agricultural regions, snow is the leading source of irrigation and drinking water. In the southwestern United States, for instance, the Colorado River—the region's primary source of irrigation water—depends on snowfields in the Rockies for much of its flow. California, in addition to depending heavily on the Colorado, also relies on snowmelt from the Sierra Nevada mountain range to supply irrigation water to the Central Valley, the country's fruit and vegetable basket.

A preliminary analysis of rising temperature effects on three major river systems in the western United States—the Columbia, the Sacramento, and the Colorado—indicates that the winter snow pack in the mountains feeding them will be reduced dramatically and that winter rainfall and flooding will increase. With a business-as-usual energy policy, global climate models project a 70-percent reduction in the snow pack for the western United States by mid-century. A detailed study of the Yakima River Valley, a vast fruit-growing region in Washington State, shows progressively heavier harvest losses as the snow pack shrinks, reducing irrigation water flows.

Agriculture in the Central Asian countries of Afghanistan, Kazakhstan, Kyrgyzstan, Tajikistan, Turkmenistan, and Uzbekistan depends heavily on snowmelt from the Hindu Kush, Pamir, and Tien Shan Mountain ranges for irrigation water. And nearby Iran gets much of its water from the snowmelt in the 5,700-meter-high Alborz Mountains between Tehran and the Caspian Sea.

The continuing loss of mountain glaciers and the reduced runoff that comes from that loss could create unprecedented water shortages and political instability in some of the world's more densely populated countries. For China, a country already struggling to contain food price inflation, there may well be spreading social unrest if food supplies tighten.

For Americans, the melting of the glaciers on the Tibetan Plateau would appear to be China's problem. It is. But it is also everyone else's problem. For U.S. consumers, this melting poses a nightmare scenario. If China enters the world market for massive quantities of grain, as it has already done for soybeans over the last decade, it will necessarily come to the United States—far and away the leading grain exporter. The prospect of 1.3 billion Chinese with rapidly rising incomes competing with American consumers for the U.S. grain harvest, and thus driving up food prices, is not an attractive one.

In the 1970s, when tight world food supplies were generating unacceptable food price inflation in the United States, the government restricted grain exports. This, however, may not be an option where China is concerned. Each month when the Treasury Department auctions off securities to cover the U.S. fiscal deficit, China is one of the big buyers. Now holding close to $900 billion of U.S. debt, China has become the banker for the United States. Like it or not, American consumers will be sharing the

U.S. grain harvest with Chinese consumers. The idea that shrinking glaciers on the Tibetan Plateau could one day drive up food prices at U.S. supermarket checkout counters is yet another sign of the complexity of our world.

Ironically, the two countries that are planning to build most of the new coal-fired power plants, China and India, are precisely the ones whose food security is most massively threatened by the carbon emitted from burning coal. It is now in their interest to try and save their mountain glaciers by quickly shifting energy investment from coal-fired power plants into energy efficiency, wind farms, solar thermal power plants, and geothermal power plants.

We know from studying earlier civilizations that declined and collapsed that shrinking harvests often were responsible. For the Sumerians, it was rising salt concentrations in the soil that lowered wheat and barley yields and eventually brought down this remarkable early civilization. For us, it is rising carbon dioxide concentrations in the atmosphere that are raising the global temperature, which ultimately could shrink grain harvests and bring down our global civilization.

Data, endnotes, and additional resources can be found on Earth Policy's Web site, at www.earth-policy.org.

II
THE CONSEQUENCES

5

The Emerging Politics of Food Scarcity

Between early 2007 and 2008, world wheat, rice, corn, and soybean prices climbed to roughly triple their historical levels. With food prices soaring, the social order in many countries began to break down. In several provinces in Thailand, rice rustlers stole grain by harvesting ripe fields during the night. In response, Thai villagers with distant rice fields took to guarding them at night with loaded shotguns.

In Sudan, the U.N. World Food Programme (WFP), the provider of grain for 2 million people in Darfur refugee camps, faced a difficult mission. During the first three months of 2008, some 56 grain-laden trucks were hijacked. The hunger relief effort itself broke down. In Pakistan, where flour prices had doubled, food security became a national concern. Thousands of armed Pakistani troops were assigned to guard grain elevators and trucks hauling wheat.

As more and more people were trapped between low incomes and rising food prices, food riots became commonplace. In Egypt, soldiers were conscripted to bake bread. Bread lines at state-subsidized bakeries were often the scene of fights and sometimes deaths. In Morocco, 34 food rioters were jailed. In Yemen, food riots turned deadly, taking at least a dozen lives. In Cameroon, dozens of people died in food riots and hundreds were arrested. Other countries where riots erupted include Ethiopia, Haiti, Indonesia, Mexico, the Philippines, and Senegal. Haiti was hit particularly hard. After a week of riots and violence, the prime minister was forced to step down.

The tripling of world grain prices also sharply reduced food aid supplies, putting the dozens of countries dependent on the WFP's emergency food assistance at risk. In March 2008, the WFP issued an urgent appeal for $500 million of additional funds. Even before the price hikes, the WFP estimated that 18,000 children were dying daily of hunger and related illnesses.

The world has experienced several grain price surges over the last half-century, but none like the one in 2007–08. The earlier surges were event-driven—a monsoon failure in India, a severe drought in the Soviet Union, or a crop-shrinking heat wave in the U.S. Midwest. The price surges were temporary, caused by weather-related events that were typically remedied by the next harvest. The record 2007–08 surge in grain prices was different. It was driven by converging trends on both sides of the food-population equation—some long-standing, others more recent.

Today there are three sources of growing demand for food: population growth; rising affluence and the associated jump in meat, milk, and egg consumption; and the use of grain to produce fuel for cars. Population growth is as old as agriculture itself. But the world is now adding close to 80 million people per year. Even worse, the overwhelming majority of these people are being added in countries where cropland is scarce, soils are eroding, and irrigation wells are going dry.

Even as we are multiplying in number, some 3 billion of us are trying to move up the food chain, consuming more grain-intensive livestock products. As incomes rise, annual grain consumption per person climbs from less than 400 pounds, as in India today, to roughly 1,600 pounds, as in the United States, where diets tend to be heavy with meat and dairy products.

The third source of demand growth emerged when the United States attempted to reduce its oil insecurity by converting grain into ethanol. The jump in U.S. gasoline prices to $3 per gallon that followed Hurricane Katrina in 2005 made it highly profitable to invest in ethanol distilleries in the United States. As a result, the growth in world grain demand, traditionally around 20 million tons per year, suddenly jumped to over 50 million tons in 2007 and again in 2008 as a huge fleet of new ethanol distilleries came online. This massive ethanol distillery investment in the United States launched an epic competition between cars and people for grain.

The conversion of grain to automotive fuel has continued to climb. Roughly 119 million tons of the 2009 U.S. grain harvest of 416 million tons went to ethanol distilleries, an amount that exceeds the grain harvests of Canada and Australia combined.

Even as these three sources of demand combined to drive up world consumption, speculators entered the fray. By buying grain futures and holding grain off the market, they further fueled the price rise.

On the supply side of the food equation, several trends discussed in preceding chapters are making it more difficult to expand production rapidly enough to keep up with demand. These include soil erosion, aquifer depletion, more-frequent crop-shrinking heat waves, melting ice sheets, melting mountain glaciers, and the diversion of irrigation water to cities.

Farmers are also losing cropland to nonfarm uses. Cars compete with people not only for the grain supply but also for the cropland itself. The United States, for example, has paved an area for cars larger than the state of Georgia. Every five cars added to the U.S. fleet means another acre of land will be paved over—the equivalent of a football field.

The implications for China of this relationship between cars and cropland are startling. In 2009, for the first time, more cars were sold in China than in the United States. If China were to reach the U.S. ownership rate of three cars for every four people, it would have over a billion cars, more than the entire world has today. The land that would have to be paved to accommodate these cars would be two thirds the area China currently has in rice.

This pressure on cropland worldwide is running up against increased demand for soybeans, which are the key to expanding the production of meat, milk, and eggs. Adding soybean meal to livestock and poultry feed sharply boosts the efficiency with which grain is converted into animal protein. This is why world soybean use climbed from 17 million tons in 1950 to 252 million tons in 2010, a 15-fold jump.

Nowhere is the soaring demand for soybeans more evident than in China, where the crop originated. As recently as 1995, China produced 14 million tons of soybeans and consumed 14 million tons. In 2010, it still produced 14 million tons, but it consumed a staggering 64 million tons. In fact, over half of the world's soybean exports now go to China.

Demand is climbing, but since scientists have failed to increase yields rapidly, the world gets more soybeans largely by planting more soybeans. The soybean is devouring land in the United States, Brazil, and Argentina, which together account for four fifths of world soybean production and 90 percent of exports. The United States now has more land in soybeans than in wheat. In Brazil, there is more land in soybeans than in corn, wheat, and rice combined. Argentina's soybean area is now double that in all grains combined. It is a virtual soybean monoculture. Soaring world demand for soybeans is thus driving deforestation in Brazil and the plowing of grasslands in Argentina.

The trends generating food demand and restricting supply are converging to create a perfect storm in the world food economy, one that is generating a new politics of food scarcity. Faced with potential domestic political instability as food prices soared, beginning in late 2007 Russia and Argentina limited or banned exports of wheat in an attempt to check domestic food price rises. Viet Nam, the number two rice exporter, banned rice exports for several months. While these moves reassured people living in the exporting countries, they created panic in the scores of countries that import grain. Governments of importing countries suddenly realized that they could no longer rely on the world market for supplies.

In response, some countries tried to nail down long-term bilateral trade agreements that would lock up future grain supplies. The Philippines, a leading rice importer, negotiated a three-year deal with Viet Nam for a guaranteed 1.5 million tons of rice each year. A delegation from Yemen traveled to Australia with the hope of negotiating a long-term wheat import deal. They failed. Other importing countries sought similar arrangements, but in a seller's market, few were successful.

The loss of confidence among importing countries has led the more affluent ones to buy or lease large blocks of land in other countries on which to produce food for themselves. In the language of the diplomatic and investment communities, these are land acquisitions. In the language of the small farmers displaced from their land and the nongovernmental organizations (NGOs) that work with them, they are land grabs.

As food supplies tighten, we are witnessing an unprecedented scramble for land that crosses national boundaries. Initially driven by food insecurity at the national level, land acquisitions are now also seen as a lucrative investment opportunity. Fatou Mbaye of ActionAid in Senegal observes, "Land is quickly becoming the new gold and right now the rush is on."

Among the countries that are leading the charge to buy or lease land abroad are Saudi Arabia, South Korea, and China. Saudi Arabia, which is fast losing its irrigation water, will soon be totally dependent on imports or overseas projects for its grain. South Korea now imports over 70 percent of its grain. China, faced with aquifer depletion and the heavy loss of cropland to nonfarm uses, is nervous. Although essentially self-sufficient in grain for over a decade, in 2010 it started to import wheat from Australia, Kazakhstan, Canada, and the United States. It also imported U.S. corn.

India, though not an affluent country, has also become a major player in land acquisitions. With its irrigation wells starting to go dry and with growing climate instability, India too is worried about future food security. Among the other countries jumping in to buy land abroad are Egypt, Libya, Bahrain, Qatar, and the United Arab Emirates.

The initial land acquisitions typically began as negotiations by governments concerned about food security. It was an interesting combination of diplomacy and business—with governments often negotiating side by side with corporations from their own countries, some formed precisely to produce food abroad. Once the negotiations are completed, the corporations usually take over. Over time, the land acquisitions have also become investment opportunities for agribusiness firms, investment banks, and sovereign wealth funds.

In Asia, the countries selling or leasing land include Indonesia, the Philippines, and Papua New Guinea. In Latin America, it is mostly Brazil, but also Argentina and Paraguay. In Africa, where land values are low compared with those in Asia, Ethiopia, Sudan, and Mozambique are among the many countries recently targeted by investors. In Ethiopia, for example, an acre of land can be leased for less than $1 per year, whereas in land-scarce Asia it could easily cost $100 or more. For land acquisitions, Africa is the new frontier.

Thus the countries selling or leasing their land are often poor and, more often than not, those where hunger is chronic, such as Ethiopia and Sudan. In January 2009 the Saudis celebrated the arrival of the first shipment of rice produced on land they had acquired in Ethiopia, a country where the WFP is currently feeding 5 million people. And Sudan is the site of the WFP's largest famine relief effort.

The purpose of land acquisition varies. For some, it is to produce food grains—rice and wheat. For others, it is to produce livestock and poultry feed, principally corn. A third factor driving land acquisitions is the demand for automotive fuel. The European Union's goal of obtaining 10 percent of its transport energy from renewable sources by 2020 is encouraging land grabbers to produce biofuels for the European market.

For sheer size of acquisitions, China stands out. The Chinese reportedly picked up 7 million acres in the Democratic Republic of the Congo (DRC) to produce palm oil, which can be used for food or fuel. Compare that with the 3 million acres used in the DRC to produce corn, the leading grain consumed by its 68 million people. Like Ethiopia and Sudan, the DRC depends on a WFP lifeline. China is also negotiating for 5 million acres in Zambia to produce jatropha, an oilseed-bearing perennial. Other countries where China has acquired land or is planning to do so include Australia, Russia, Brazil, Kazakhstan, Myanmar, and Mozambique.

South Korea, a leading importer of corn and wheat, is a major land investor in several countries. With deals signed for 1.7 million acres in Sudan for growing wheat, South Korea is a leader in this food security push. For perspective, this is not much smaller than the 2.3 million acres South Korea uses at home to produce rice, a crop in which it is self-sufficient. Saudi Arabia is acquiring land in Ethiopia, Sudan, Indonesia, and the Philippines, while India's early investments are concentrated in several African countries, although principally in Ethiopia.

One of the little noticed characteristics of land acquisitions is that they are also water acquisitions. Whether the land is irrigated or rain-fed, it represents a claim on the water resources in the host country. This means land acquisition agreements are an even more sensitive issue in water-stressed countries. Land acquisitions in Ethiopia, where most of the Nile's headwaters begin, or in Sudan, which taps water from the Nile downstream, may simply mean that Egypt will get less of the river's water—pushing its heavy dependence on imported grain even higher.

Another disturbing dimension of many land investments is that they are taking place in countries like Indonesia, Brazil, and the DRC where expanding cropland often means clearing tropical rainforests that sequester carbon. Land clearing here could markedly raise global carbon emissions, further increasing climate change's disruptive effect on food security.

Bilateral land acquisitions raise many questions. To begin with, these agreements are almost always negotiated in secret. Typically only a few high-ranking officials are involved and the terms are often kept confidential, even though they deal with land, a public resource. Not only are key stakeholders such as local farmers not at the negotiating table, they often do not even learn about the agreements until after the papers are signed. And since there is rarely productive land sitting idle in the countries where the land is being acquired, the agreements mean that many local farmers and herders will simply be displaced. Their land may be confiscated or it may be bought from them at a price over which they have little say, leading to the public hostility that often arises in host countries.

In a landmark article on the African land grab in *The Observer*, John Vidal quotes an Ethiopian, Nyikaw Ochalla, from the Gambella region: "The foreign companies are arriving in large numbers, depriving people of land they have used for centuries. There is no consultation with the indigenous population. The deals are done secretly. The only thing the local people see is people coming with lots of tractors to invade their lands." Referring to his own village, where an Indian corporation is taking over, Ochalla says, "Their land has been compulsorily taken and they have been given no compensation. People cannot believe what is happening."

Hostility of local people to land grabs is the rule, not the exception. China, for example, signed an agreement with the Philippine government in 2007 to lease 2.5 million acres of land on which to produce crops that would be shipped home. Once word leaked out, the public outcry—much of it from Filipino farmers—forced the government to suspend the agreement. A similar situation developed in Madagascar, where a South Korean firm, Daewoo Logistics, had pursued rights to more than 3 million acres of land, an area half the size of Belgium. This helped stoke a political furor that led to a change in government and cancellation of the agreement.

Investments by agribusiness firms and others to acquire land in low-income countries and to produce food exclusively for export are almost certainly going to leave people in these countries less well off. Many will be left landless. At the national level, there will be less land to produce food for local use.

If food prices are rising in the host country, will the investing country actually be able to remove the grain it has produced on acquired land? Will the hungry people in these countries stand by and watch as grain is exported from land that was once theirs? Or will the investors have to hire security forces to ensure that the harvests can be shipped home? Those acquiring land in hungry countries are sowing what could become the seeds of conflict.

The central question associated with this massive effort by importing countries to acquire land abroad is this: How will it affect world food production and overall food security? In a September 2010 report, the World Bank used press reports to identify 464 land acquisitions that were in various stages of development between October 2008 and August 2009. The Bank claimed that production had begun on only one fifth of the announced projects, partly because many deals were made by land speculators. The report offered several other reasons for the slow start, including "unrealistic objectives, price changes, and inadequate infrastructure, technology, and institutions."

The land area involved was clear for only 203 of these reported projects, yet it still came to some 115 million acres, an area comparable to the U.S. land in corn and wheat combined. These agreements imply an investment of at least $50 billion. Particularly noteworthy is that of the 405 projects for which commodity information was available, 21 percent are slated to produce biofuels—and another 21 percent industrial or cash crops. Only 37 percent are slated to produce food crops.

How productive will the land be that actually ends up being farmed? Given the level of agricultural skills and technologies likely to be used, in most cases relatively high yields can be expected. In Africa, for example, simply applying fertilizer to its nutrient-depleted soils will often double grain yields. Taking everything into account, investors should be able to double or triple yields in much of Africa.

While there will undoubtedly be some spectacular production gains in some countries with some crops, there will also be occasional failures. Some projects will be abandoned because the economics simply do not work. Long-distance farming, with the transportation and travel involved, and at a time when oil prices are likely to be rising, can be very costly. There almost certainly will be unforeseen outbreaks of plant disease and insect infestations as new crops are introduced, particularly since so much land acquired is in tropical and subtropical regions.

Another uncertainty is the timing. As the Bank study indicates, all of this land will not automatically come into production within a year or two. Although the flurry of reports of large-scale land acquisitions began in 2008, as of 2010 there were only a few small harvests to point to. As noted, the Saudis harvested their first rice crop in Ethiopia in late 2008. In 2009, South Korea's Hyundai Heavy Industries harvested some 4,500 tons of soybeans and 2,000 tons of corn on a 25,000-acre farm it took over from Russian owners, roughly 100 miles north of Vladivostok. Hyundai plans to expand this to 125,000 acres by 2012, and by 2015 it expects to produce 100,000 tons of soybeans and corn annually for the Korean market, less than 1 percent of South Korea's consumption of these two commodities. And an Indian firm has started harvesting corn in Ethiopia.

The public infrastructure to support modern market-oriented agriculture does not yet exist in much of Africa. In some countries, it will take years to build the roads needed both to bring in agricultural inputs, such as fertilizer, and to export the farm products. Modern agriculture requires its own infrastructure—machine sheds, grain silos, fertilizer storage sheds, fuel storage facilities, and, in many situations, irrigation pumps and well-drilling equipment. Overall, land development to date appears to be a slow, time-consuming process.

Even if some of these projects can dramatically boost land productivity, there is also the question of whether local people will benefit. If virtually all the inputs—the farm equipment, the fertilizer, the pesticides, the seeds—are brought in from abroad and if all the output is shipped out of the country, it will not contribute to the local economy or the local food supply. At best, people from local communities may get work as farm laborers, but in highly mechanized operations, jobs will be few. At worst, countries will be left with less land and water with which to feed their already hungry populations.

One of the most difficult variables to evaluate is political stability. Once opposition political parties are in office, they may cancel the agreements, arguing that they were secretly negotiated without public participation or support. Land acquisitions in the DRC and Sudan, both among the top five failing states, are particularly risky. Few things are more likely to fuel insurgencies than taking land away from people. Agricultural equipment is easily sabotaged. If ripe fields of grain are torched, they burn quickly.

The World Bank, working with the U.N. Food and Agriculture Organization and other related agencies, has formulated a set of principles governing land acquisitions. These guiding principles are well conceived, but there is no mechanism to enforce them. The Bank does not seem eager to challenge the basic argument of those acquiring land, namely that it will benefit those who live in the host countries.

But the land acquisitions are being fundamentally challenged by a coalition of more than 100 NGOs, some national and others international. These groups argue that what the world needs is not large corporations bringing large-scale, highly mechanized, capital-intensive agriculture into these countries, but international support for community-based farming, centered around labor-intensive family farms that produce for local and regional markets and that create desperately needed jobs.

As land and water become scarce, as the earth's temperature rises, and as world food security deteriorates, a dangerous geopolitics of food scarcity is emerging. The conditions giving rise to this have been in the making for several decades, but the situation has come into sharp focus only in the last few years. Land grabbing is an integral part of a global power struggle for food security. Not only is it designed to benefit the rich, it will likely do so at the expense of the poor.

Data, endnotes, and additional resources can be found on Earth Policy's Web site, at www.earth-policy.org.

6

Environmental Refugees: The Rising Tide

In late August 2005, as Hurricane Katrina approached the U.S. Gulf Coast, more than 1 million people were evacuated from New Orleans and the small towns and rural communities along the coast. The decision to evacuate was well taken. In some Gulf Coast towns, Katrina's powerful 28-foot-high storm surge did not leave a single structure standing. New Orleans survived the initial hit, but it was flooded when the inland levies were breached and water covered large parts of the city—in many cases leaving just the rooftops exposed, where thousands of people were stranded.

Once the storm passed, it was assumed that the million or so Katrina evacuees would, as in past cases, return to repair and rebuild their homes. Some 700,000 did return, but close to 300,000 did not. Nor do they plan to do so. Most of them have no home or job to return to. They are no longer evacuees. They are climate refugees. Interestingly, the first large wave of modern climate refugees emerged in the United States—the country most responsible for the rise in atmospheric carbon dioxide that is warming the earth. New Orleans is the first modern coastal city to be partly abandoned.

One of the defining characteristics of our time is the swelling flow of environmental refugees: people displaced by rising seas, more-destructive storms, expanding deserts, water shortages, and dangerously high levels of toxic pollutants in the local environment.

Over the longer term, rising-sea refugees will likely dominate the flow of environmental refugees. The prospect for this century is a rise in sea level of up to 6 feet. Even a 3-foot rise would inundate parts of many low-lying cities, major river deltas, and low-lying island countries. Among the early refugees will be millions of rice-farming families from Asia's low-lying river deltas, those who will watch their fields sink below the rising sea.

The flow of rising-sea refugees will come primarily from coastal cities. Among those most immediately affected are London, New York, Washington, Miami, Shanghai, Kolkata (Calcutta), Cairo, and Tokyo. If the rise in sea level cannot be checked, cities soon will have to start either planning for relocation or building barriers that will block the rising seas.

The movement of millions of rising-sea refugees to higher elevations in the interior of their countries will create two real estate markets—one in coastal regions, where prices will fall, and another in the higher elevations, where they will rise. Property insurance rates are already rising in storm- and flood-prone places like Florida.

River deltas contain some of the largest, most vulnerable populations. These include the deltas of the Mekong, Irrawaddy, Niger, Nile, Mississippi, Ganges-Brahmaputra, and Yangtze Rivers. For example, a 6-foot sea level rise would displace 15 million Bangladeshis living in the densely populated Ganges-Brahmaputra delta.

The London-based Environmental Justice Foundation reports that "a one meter [3 foot] sea-level rise would affect up to 70 percent of Nigeria's coastline affecting over 2.7 million hectares. Egypt would lose at least 2 million hectares in the fertile Nile Delta, displacing 8 to 10 million people, including nearly the entire population of Alexandria."

Low-lying islands will also be hit hard. The 39 members of the Alliance of Small Island States stand to lose part or all of their territories as sea level rises. Among the most immediately threatened are Tuvalu, Kiribati, and the Marshall Islands in the Pacific Ocean and the Maldives in the Indian Ocean. Well before total inundation, islanders face salt water intrusion that can contaminate their drinking water and make it impossible for deep-rooted crops to survive. Eventually, all crops will fail.

Some 3,000 of Tuvalu's 10,000 people have already migrated to New Zealand, seeking work under a labor migration program. Larger populations, such as the 300,000 people in the Maldives, will find it more difficult to migrate elsewhere. The president of the Maldives is actively pursuing the possibility of purchasing land for his people to migrate to as the sea level inches upward and makes island life untenable.

Meanwhile, following the 2004 tsunami that so memorably devastated Indonesia, the government of the Maldives decided to organize a "staged retreat" by moving people from the lower-lying islands, some 200 in total, to a dozen or so slightly higher islands. But even the highest of these is only about 8 feet above sea level. And in anticipation of higher seas, the Papua New Guinea government moved the 1,000 residents of the Carteret Islands to the larger island of Bougainville.

Aside from the social upheaval and the personal devastation of people losing their country to the rising sea, there are also legal issues to be resolved. When does a country cease to exist legally, for example? Is it when there is no longer a functioning government? Or when it has disappeared beneath the waves? And at what point does a country lose its vote in the United Nations? In any event, rising sea level is likely to shrink U.N. membership as low-lying island states disappear.

How far might the sea level rise? Rob Young and Orrin Pilkey note in *The Rising Sea* that planning panels in Rhode Island and Miami assume a minimum rise of 3.5 feet by 2100. A California planning study uses a 4.6-foot rise by century's end. The Dutch, for their coastal planning purposes, are assuming a 2.5-foot rise for 2050.

If the Greenland ice sheet, which is well over a mile thick in places, were to melt completely, sea level would rise 23 feet. And if the West Antarctic ice sheet were to break up entirely, sea level would rise 16 feet. Together, the melting of these two ice sheets, which scientists believe to be the most vulnerable, would raise sea level 39 feet. And this does not include thermal expansion as ocean water warms, an important contributor to sea level rise.

A study published by the International Institute for Environment and Development has analyzed the effect of a 10-meter (33-foot) rise in sea level. The study begins by noting that 634 million people currently live along coasts at 10 meters or less above sea level, in what they call the Low Elevation Coastal Zone.

The most vulnerable country is China, with 144 million potential climate refugees. India and Bangladesh are next, with 63 million and 62 million respectively. Viet Nam has 43 million vulnerable people, and Indonesia 42 million. Also in the top 10 are Japan with 30 million, Egypt with 26 million, and the United States with 23 million. Some of the refugees could simply retreat to higher ground within their own country. Others—facing extreme crowding in the interior regions of their homeland—would seek refuge elsewhere.

The second category of environmental refugees is also closely related to elevated global temperatures. A higher surface water temperature in the tropical oceans means there is more energy to drive tropical storm systems, which can lead to more-destructive storms. The combination of more-powerful storms and stronger storm surges can be devastating, as New Orleans discovered. The regions that are most at risk for more-powerful and destructive storms are Central America, the Caribbean, and both the Atlantic and Gulf coasts of the United States. In Asia, where hurricanes are called typhoons, it is East and Southeast Asia, including Japan, China, Taiwan, the Philippines, and Viet Nam, that are most vulnerable. The other region in danger is the Bay of Bengal, particularly Bangladesh.

In the fall of 1998, Hurricane Mitch—one of the most powerful storms ever to come out of the Atlantic, with winds approaching 200 miles per hour—hit the east coast of Central America. As atmospheric conditions stalled the normal northward progression of the storm, more than 6 feet of rain fell on parts of Honduras and Nicaragua within a few days. The deluge collapsed homes, factories, and schools, leaving them in ruins. It destroyed roads and bridges. Seventy percent of the crops in Honduras were washed away, as was much of the topsoil. Huge mudslides destroyed villages, sometimes burying local populations.

The storm left 11,000 dead. Thousands more were never found. The basic infrastructure—the roads and bridges in Honduras and Nicaragua—was largely destroyed. President Flores of Honduras summed it up this way: "Overall, what was destroyed over several days took us 50 years to build." The cost of the damage from this storm exceeded the annual gross domestic product of the two countries and set their economic development back by 20 years.

The first decade of this century has brought many other destructive storms. In 2004, Japan experienced a record 10 typhoons that collectively caused $10 billion worth of losses. The 2005 Atlantic hurricane season was the worst on record, bringing 15 hurricanes, including Katrina, and $115 billion in insured losses.

A third source of refugees is advancing deserts, which are now on the move almost everywhere. The Sahara desert is expanding in every direction. As it advances northward, it is squeezing the populations of Morocco, Tunisia, and Algeria against the Mediterranean coast.

The Sahelian region of Africa—the vast swath of savannah that separates the southern Sahara desert from the tropical rainforests of central Africa—is shrinking as the desert moves southward. As the desert invades Nigeria, Africa's most populous country, from the north, farmers and herders are forced southward, squeezed into a shrinking area of productive land. Some desert refugees end up in cities, many in squatter settlements, others migrate abroad. A 2006 U.N. conference on desertification in Tunisia projected that by 2020 up to 60 million people could migrate from sub-Saharan Africa to North Africa and Europe.

In Iran, villages abandoned because of spreading deserts or a lack of water number in the thousands. In the vicinity of Damavand, a small town within an hour's drive of Tehran, 88 villages have been abandoned.

In Latin America, expanding deserts are forcing people to move in both Brazil and Mexico. In Brazil, some 250,000 square miles of land are affected, much of it concentrated in the country's northeast. In Mexico, many of the migrants who leave rural communities in arid and semiarid regions of the country each year are doing so because of desertification. Some of these environmental refugees end up in Mexican cities, others cross the northern border into the United States. U.S. analysts estimate that Mexico is forced to abandon 400 square miles of farmland to desertification each year.

In China, desert expansion has accelerated in each successive decade since 1950. Desert scholar Wang Tao reports that over the last half-century or so some 24,000 villages in northern and western China have been abandoned either entirely or partly because of desert expansion.

China's Environmental Protection Agency reports that from 1994 to 1999 the Gobi Desert grew by 20,240 square miles, an area half the size of Pennsylvania. With the advancing Gobi now within 150 miles of Beijing, China's leaders are beginning to sense the gravity of the situation.

The U.S. Dust Bowl of the 1930s, which was caused by overplowing and triggered by drought, forced more than 2 million "Okies" to leave the land, many of them heading west from Oklahoma, Texas, and Kansas to California. But the dust bowl forming in China is much larger and so is the population: during the 1930s the U.S. population was only 150 million, compared with China's 1.3 billion today. Whereas U.S. migration was measured in the millions, China's may measure in the tens of millions. And as a U.S. embassy report entitled *Grapes of Wrath in Inner Mongolia* noted, "unfortunately, China's twenty-first century 'Okies' have no California to escape to—at least not in China."

The fourth group of people who will be forced to leave their homes are those in places where water tables are falling. With the vast majority of the 3 billion people projected to be added to the world by 2050 being born in such countries, water refugees are likely to become commonplace. They will be most common in arid and semiarid regions where populations are outgrowing the water supply and sinking into hydrological poverty. Villages in northwestern India are being abandoned as aquifers are depleted and people can no longer find water. Millions of villagers in northern and western China and in northern Mexico may have to move because of a lack of water.

Thus far the evacuations resulting from water shortages have been confined to villages, but eventually whole cities might have to be relocated, such as Sana'a, the capital of Yemen, and Quetta, the capital of Pakistan's Baluchistan province. Sana'a, a fast-growing city of more than 2 million people, is literally running out of water. Wells that are 1,300 feet deep are beginning to go dry. In this "race to the bottom" in the Sana'a valley, oil drilling equipment is being used to dig ever deeper wells. Some are now over half a mile deep.

The situation is bleak because trying to import water into this mountain valley from other provinces would generate tribal conflicts. Desalting sea water on the coast would be expensive because of the cost of the process itself, the distance the water would have to be pumped, and the city's altitude of 7,000 feet. Sana'a may soon be a ghost city.

Quetta, originally designed for 50,000 people, now has a population exceeding 1 million, all of whom depend on 2,000 wells pumping water from what is believed to be a fossil aquifer. In the words of one study assessing its water prospect, Quetta will soon be "a dead city."

Two other semiarid Middle Eastern countries that are suffering from water shortages are Syria and Iraq. Both are beginning to reap the consequences of overpumping their aquifers, namely irrigation wells going dry.

In Syria, these trends have forced the abandonment of 160 villages. Hundreds of thousands of farmers and herders have left the land and pitched tents on the outskirts of cities, hoping to find work. A U.N. report estimates that more than 100,000 people in northern Iraq have been uprooted because of water shortages. Hussein Amery, a Middle East water expert from the Colorado School of Mines, puts it very simply: "Water scarcity is forcing people off the land."

The fifth category of environmental refugee has

appeared only in the last 50 years or so: people who are trying to escape toxic waste or dangerous radiation levels. During the late 1970s, Love Canal—a small town in upstate New York, part of which was built on top of a toxic waste disposal site—made national and international headlines. Beginning in 1942, the Hooker Chemical Company had dumped 21,000 tons of toxic waste, including chlorobenzene, dioxin, halogenated organics, and pesticides there. In 1952, Hooker closed the site, capped it over, and deeded it to the Love Canal Board of Education. An elementary school was built on the site, taking advantage of the free land.

But during the 1960s and 1970s people began noticing odors and residues from seeping wastes. Birth defects and other illnesses were common. Beginning in August 1978, families were relocated at government expense and reimbursed for their homes at market prices. By October 1980, a total of 950 families had been permanently relocated.

A few years later, the residents of Times Beach, Missouri, began complaining about various health problems. A firm spraying oil on roads to control dust was using waste oil laden with toxic chemical wastes. After the U.S. Environmental Protection Agency discovered dioxin levels well above the public health standards, the federal government arranged for the permanent evacuation and relocation of the town's 2,000 people.

Another infamous source of environmental refugees is the Chernobyl nuclear power plant in Kiev, which exploded in April 1986. This started a powerful fire that lasted for 10 days. Massive amounts of radioactive material were spewed into the atmosphere, showering communities in the region with heavy doses of radiation. As a result, the residents of the nearby town of Pripyat and several other communities in Ukraine, Belarus, and Russia were evacuated, requiring the resettlement of 350,400 people. In 1992, six years after the accident, Belarus was devoting 20 percent of its national budget to resettlement and the many other costs associated with the accident.

While the United States has relocated two communities because of health-damaging pollutants, the identification of 459 "cancer villages" in China suggests the need to evacuate hundreds of communities. China's Ministry of Health statistics show that cancer is now the country's leading cause of death. The lung cancer death rate, also boosted by smoking, has risen nearly fivefold over the last 30 years.

With little pollution control, whole communities near chemical factories are suffering from unprecedented rates of cancer. The World Bank reports that liver cancer death rates among China's rural population are four times the global average. Their stomach cancer death rates are double those for the world. Chinese industrialists build factories in rural areas where there is cheap labor and little or no enforcement of pollution control laws. Young people are leaving for the city in droves, for jobs and possibly for better health. Yet many others are too sick or too poor to leave.

Separating out the geneses of today's refugees is not always easy. Often the environmental and economic stresses that drive migration are closely intertwined. But whatever the reason for leaving home, people are taking increasingly desperate measures. The news headlines about refugees who try to cross the Mediterranean tell the story: a 2009 BBC story entitled "Hundreds Feared Drowned off Libya," a 2008 *Guardian* piece with the headline "Over 70 Migrants Feared Killed on Crossing to Europe," and an Associated Press story from 2008—"Spain: 35 Reported Dead in Migrant Ordeal."

Some of the stories are heartrending beyond belief. In mid-October 2003, Italian authorities discovered a boat bound for Italy carrying refugees from Africa. After being adrift for more than two weeks and having run out of fuel, food, and water, many of the passengers had died. At first the dead were tossed overboard. But after a point, the remaining survivors lacked the strength to hoist the bodies over the side. The dead and the living shared the boat, resembling what a rescuer described as "a scene from Dante's Inferno."

The refugees were believed to be Somalis who had embarked from Libya, but the survivors would not reveal their country of origin lest they be sent home. We do not know whether they were political, economic, or environmental refugees. Failed states like Somalia produce all three. We do know that Somalia is a lawless entity and an ecological basket case, with overpopulation, overgrazing, and the resulting desertification destroying its pastoral economy.

In April 2006, a man fishing off the coast of Barbados discovered a 20-foot boat adrift with the bodies of 11 young men on board, bodies that were "virtually mummified" by the sun and salty ocean spray. As the end drew near, one passenger left a note tucked between two bodies: "I would like to send my family in Basada [Senegal] a sum of money. Please excuse me and goodbye." The author of the note was apparently one of a group of 52 who had left Senegal on Christmas Eve aboard a boat destined for the Canary Islands, a jumping off point for Europe.

Each day Mexicans risk their lives in the Arizona desert, trying to reach jobs in the United States. Some 400–600 Mexicans leave rural areas every day, abandoning plots of land too small or too eroded to make a living. They either head for Mexican cities or try to cross illegally into the United States. Many of those who try to cross the Arizona desert perish in its punishing heat; scores of bodies are found along the Arizona border each year.

The potentially massive movement of people across national boundaries is already affecting some countries. India, for example, with a steady stream of migrants from Bangladesh and the prospect of millions more to come, is building a 10-foot-high fence along their shared border. The United States is erecting a fence along the border with Mexico. The current movement of Chinese across the border into Siberia is described as temporary, but it will likely become permanent. Another major border, the Mediterranean Sea, is now routinely patrolled by naval vessels trying to intercept the small boats of African migrants bound for Europe.

In the end, the question is whether governments are strong enough to withstand the political and economic stress of extensive migration flows, both internal and external. Some of the largest flows will be across national borders and they are likely to be illegal. As a general matter, environmental refugees will be migrating from poor countries to rich ones, from Africa, Asia, and Latin America to North America and Europe. In the face of mounting environmental stresses, will the migration of people be limited and organized or will it be massive and chaotic?

People do not normally leave their homes, their families, and their communities unless they have no other option. Maybe it is time for governments to consider whether it might not be cheaper and far less painful in human terms to treat the causes of migration rather than merely respond to it. This means working with developing countries to restore their economy's natural support systems—the soils, the grasslands, the forests—and it means accelerating the shift to smaller families to help people break out of poverty. Treating symptoms instead of causes is not good medicine. Nor is it good public policy.

7

Mounting Stresses, Failing States

In late November 2009, Somali pirates captured a Greek-owned supertanker, the *Maran Centaurus*, in the Indian Ocean. Carrying 2 million barrels of oil, the ship's cargo was valued at more than $150 million. After nearly two months of negotiations, a $7 million ransom was paid—$5.5 million in cash was dropped from a helicopter on to the deck of the *Centaurus*, and $1.5 million was deposited in a private bank account.

This modern version of piracy in the high seas is dangerous, disruptive, costly, and amazingly successful. In an effort to stamp it out, some 17 countries—including the United States, France, Russia, and China—have deployed naval units in the region, but with limited success. In 2009, Somali pirates attacked 217 vessels at sea and succeeded in hijacking 47 of them, holding them for ransom. This was up from 111 ships attacked in 2008, 42 of which were captured. And because ransoms were larger in 2009, pirate "earnings" were roughly double those in 2008.

Somalia, a failed state, is now ruled by tribal leaders and jihadist groups, each claiming a piece of what was once a country. There is no functional national government. Part of the south is controlled by Al Shabab, a radical group affiliated with Al Qaeda. Now training terrorists, Al Shabab claimed credit in July 2010 for detonating bombs in two crowds that had gathered in Kampala, Uganda, to watch the World Cup soccer championship match on television. At least 70 people were killed and many more injured.

Uganda was a target because it supplied troops for an African peacekeeping force in Somalia. Al Shabab is also anti-soccer, banning both the playing and watching of this "infidel" sport in the territory it controls. Somalia is thus now both a base for pirates and a training ground for terrorists. As *The Economist* has observed, "like a severely disturbed individual, a failed state is a danger not just to itself but to those around it and beyond."

After a half-century of forming new states from former colonies and from the breakup of the Soviet Union, the international community is today faced with the opposite situation: the disintegration of states. The term "failing state" has been in use only a decade or so, but these countries are now a prominent feature of the international political landscape. As an article in *Foreign Policy* observes, "Failed states have made a remarkable odyssey from the periphery to the very center of global politics."

In the past, governments worried about the concentration of too much power in one state, as in Nazi Germany, Imperial Japan, and the Soviet Union. But today it is failing states that provide the greatest threat to global order and stability. As *Foreign Policy* notes, "World leaders once worried about who was amassing power; now they worry about the absence of it."

Some national and international organizations maintain their own lists of failing, weak, or fragile states, as they are variously called. The U.S. Central Intelligence Agency funds the Political Instability Task Force to track political risk factors. The British government's international development arm has identified 46 "fragile states." The World Bank focuses its attention on some 30 low-income "fragile and conflict-affected countries."

But the most systematic ongoing effort to analyze countries according to their vulnerability to failure is one undertaken by the Fund for Peace and published in each July/August issue of *Foreign Policy*. This invaluable annual assessment, which draws on thousands of information sources worldwide, is rich with insights into the changes that are under way in the world and, in a broad sense, where the world is heading.

The research team analyzes data for 177 countries and ranks them according to "their vulnerability to violent internal conflict and societal deterioration." It puts Somalia at the top of the 2010 Failed States Index, followed by Chad, Sudan, Zimbabwe, and the Democratic Republic of the Congo (DRC). (See Table 7–1.) Three oil-exporting countries are among the top 20: Sudan, Iraq, and Nigeria. Pakistan, now ranked at number 10, is the only failing state with a nuclear arsenal, but North Korea—nineteenth on the list—is developing a nuclear capability.

The index is based on 12 social, economic, and political indicators, including population growth, economic inequality, and legitimacy of government. Scores for each indicator, ranging from 1 to 10, are aggregated into a single country indicator. A score of 120 would mean that a society is failing totally by every measure. In the first *Foreign Policy* listing in 2005, based on data from 2004, just 7 countries had scores of 100 or more. In 2006 this increased to 9. By 2009 it was 14—doubling in four years. In 2010, it was 15. This short trend is far from definitive, but higher scores for countries at the top and the doubling of countries with scores of 100 or higher suggest that state failure is both spreading and deepening.

The most conspicuous indication of state failure is a breakdown in law and order and the related loss of personal security. States fail when national governments lose control of part or all of their territory and can no longer

Table 7–1. *Top 20 Failing States, 2010*

Rank	Country	Score
1	Somalia	114.3
2	Chad	113.3
3	Sudan	111.8
4	Zimbabwe	110.2
5	Dem. Republic of the Congo	109.9
6	Afghanistan	109.3
7	Iraq	107.3
8	Central African Republic	106.4
9	Guinea	105.0
10	Pakistan	102.5
11	Haiti	101.6
12	Côte d'Ivoire	101.2
13	Kenya	100.7
14	Nigeria	100.2
15	Yemen	100.0
16	Burma	99.4
17	Ethiopia	98.8
18	East Timor	98.2
19	North Korea	97.8
20	Niger	97.8

Source: "The Failed States Index," *Foreign Policy*, July/August 2010.

ensure people's security. When governments lose their monopoly on power, the rule of law begins to disintegrate. At this point, they often turn to the United Nations for help. In fact, 8 of the top 20 countries are being assisted by U.N. peacekeeping forces, including Haiti, Sudan, and the Democratic Republic of the Congo. The number of peacekeeping missions doubled between 2002 and 2008.

Failing states often degenerate into civil war as opposing groups vie for power. In Haiti, armed gangs ruled the streets until a U.N. peacekeeping force arrived in 2004. In Afghanistan, the local warlords or the Taliban, not the central government, control the country outside of Kabul.

One more recent reason for government breakdowns is the inability to provide food security, not necessarily because the government is less competent but because obtaining enough food is becoming more difficult. Providing sufficient food has proved to be particularly challenging since the rise in food prices that began in early 2007. Although grain prices have subsided somewhat from the peak in the spring of 2008, they are still well above historical levels. For low-income, food-deficit countries, finding enough food is becoming ever more challenging.

With food security, as with personal security, there is a U.N. fallback. The food equivalent of the peacekeeping forces is the World Food Programme (WFP), a U.N. agency providing emergency food aid in more than 60 countries, including 19 of the top 20 countries on *Foreign Policy*'s list of failing states. Some countries, such as Haiti, depend on a U.N. peacekeeping force to maintain law and order and on the WFP for part of its food. Haiti is, in effect, a ward of the United Nations.

Failing states are rarely isolated phenomena. Conflicts can easily spread to neighboring countries, as when the genocide in Rwanda spilled over into the DRC, where an ongoing civil conflict claimed more than 5 million lives between 1998 and 2007. The vast majority of these deaths in the DRC were due to war's indirect effects, including hunger, respiratory illnesses, diarrhea, and other diseases as millions of people have been driven from their homes. Similarly, the killings in Sudan's Darfur region quickly spread into Chad as victims fled across the border.

Failing states such as Afghanistan and Myanmar (Burma) have become sources of drugs. In 2009, Afghanistan supplied 89 percent of the world's opium, much of it made into heroin. Myanmar, though a distant second, is a major heroin supplier for China.

The conditions of state failure may be a long time in the making, but the collapse itself can come quickly. Yemen, for example, is facing several threatening trends. It is running out of both oil and water. The underground basin that supplies Sana'a, the capital, with water may be fully depleted by 2015. The production of oil, which accounts for 75 percent of government revenue and an even larger share of export earnings, fell by nearly 40 percent from 2003 to 2009. And with its two main oil fields seriously depleted, there is nothing in sight to reverse the decline.

Underlying these stresses is a fast-growing, poverty-stricken population, the poorest among the Arab countries, and an unemployment rate estimated at 35 percent. On the political front, the shaky Yemeni government faces a Shiite insurgency in the north, a deepening of the traditional conflict between the north and the south, and an estimated 300 Al Qaeda operatives within its borders. With its long, porous border with Saudi Arabia, Yemen could become a staging ground and a gateway for Al Qaeda to move into Saudi Arabia. Could the ultimate Al Qaeda goal of controlling Saudi Arabia, both a center of Islam and the world's leading exporter of oil, finally be within reach?

Ranking on the Failed States Index is closely linked with demographic indicators. The populations in 15 of the top 20 failing states are growing between 2 and 4 percent a year. Niger tops this list at 3.9 percent, and Afghanistan's population is growing by 3.4 percent. A population growing at 3 percent a year may not sound overwhelming, but it will expand twentyfold in a century.

In failing states, big families are the norm, not the exception, with women in a number of countries bearing an average of six or more children each.

In 14 of the top 20 failing states, at least 40 percent of the population is under 15, a demographic indicator that raises the likelihood of future political instability. Young men, lacking employment opportunities, often become disaffected and ready recruits for insurgencies.

In many of the countries with several decades of rapid population growth, governments are suffering from demographic fatigue, unable to cope with the steady shrinkage in cropland and freshwater supply per person or to build schools fast enough for the swelling ranks of children. Sudan is a classic case of a country caught in the demographic trap. Like many failing states, it has developed far enough economically and socially to reduce mortality but not far enough to lower fertility.

As a result, large families beget poverty and poverty begets large families. This is the trap. Women in Sudan have on average four children, double the number needed for replacement, expanding the population of 42 million by 2,000 per day. Under this pressure, Sudan—like scores of other countries—is breaking down.

All but 4 of the 20 countries that lead the list of failing states are caught in this demographic trap. Realistically, they probably cannot break out of it on their own. They will need outside help in raising educational levels, especially of girls. In every society for which we have data, the more education women have, the smaller their families. And the smaller families are, the easier it is to break out of poverty.

Among the top 20 countries on the 2010 Failed States list, all but a few are losing the race between food production and population growth. Even getting food relief to failing states can be a challenge. In Somalia, threats from Al Shabab and the killing of food relief workers effectively ended efforts to provide food assistance in the southern part of the hunger-stricken country.

Another characteristic of failing states is the deterioration of the economic infrastructure—roads, power, water, and sewage systems. For example, a lack of maintenance has left many irrigation canal networks built in an earlier era in an advanced state of disrepair, often no longer able to deliver water to farmers.

Virtually all of the top 20 countries are depleting their natural assets—forests, grasslands, soils, and aquifers—to sustain their rapidly growing populations. The 3 countries at the top of the list—Somalia, Chad, and Sudan—are losing their topsoil to wind erosion. The ongoing loss of topsoil is slowly undermining the land's productivity. Several countries in the top 20 are water-stressed and are overpumping their aquifers, including Afghanistan, Iraq, Pakistan, and Yemen.

After a point, as rapid population growth, deteriorating environmental support systems, and poverty reinforce each other, the resulting instability makes it difficult to attract investment from abroad. Even public assistance programs from donor countries are sometimes phased out as the security breakdown threatens the lives of aid workers. A drying up of foreign investment and an associated rise in unemployment are also part of the decline syndrome.

In an age of increasing globalization, a functioning global society depends on a cooperative network of stable nation states. When governments lose their capacity to govern, they can no longer collect taxes, much less be responsible for their international debts. More failing states mean more bad debt. Efforts to control international terrorism also depend on cooperation among functioning nation states. As more and more states fail, this cooperation becomes less and less effective.

Failing states may lack a health care system that is

sophisticated enough to participate in the international network that controls the spread of infectious diseases, such as polio, or of diseases that affect both animals and people, such as avian flu, swine flu, and mad cow disease. In 1988 the international community launched an effort to eradicate polio, a campaign patterned on the highly successful one that eliminated smallpox. The goal was to get rid of the dreaded disease that used to paralyze an average of 1,000 children each day. By 2003 polio had been eradicated in all but a few countries, among them Afghanistan, India, Nigeria, and Pakistan.

But that year mullahs in northern Nigeria began to oppose the vaccination program, claiming that it was a plot to spread AIDS and sterility. As a result, the local vaccination effort broke down, and polio cases in Nigeria tripled over the next three years. Meanwhile, Nigerian Muslims making their annual pilgrimage to Mecca may have spread the disease, reintroducing the virus in some Muslim countries, such as Indonesia, Chad, and Somalia, that were already polio-free. In response, Saudi officials imposed a polio vaccination requirement on all younger visitors from countries with reported cases of polio.

In early 2007, when eradication again appeared to be in sight, violent opposition to vaccinations arose in Pakistan's Northwest Frontier Province, where a doctor and a health worker in the Polio Eradication Program were killed. More recently, the Taliban has refused to let health officials administer polio vaccinations in the Swat Valley of Pakistan, further delaying the campaign. This raises a troubling question: In a world of failing states, is the goal of eradicating polio, once so close at hand, now beyond our reach?

Thus far, failing states have been mostly smaller ones. But some countries with over 100 million people, such as Pakistan and Nigeria, are working their way up the list. So is Mexico, where both oil production and exports have peaked, depriving the government of tax revenue and foreign exchange. Beyond this, a criminal organization called the Zetas taps government oil pipelines in areas it controls. In 2008 and 2009, it withdrew over $1 billion worth of oil. The government's war with the drug cartels has claimed 16,000 lives since 2006, a number far beyond American lives lost in Iraq and Afghanistan over the last decade. With income from oil and tourism shrinking and with foreign investors becoming nervous, the Mexican government is being seriously challenged.

For India (now number 79 on *Foreign Policy*'s list), where 15 percent of the people are being fed with grain produced by overpumping, it could be emerging water shortages translating into food shortages that triggers its decline. As local conflicts over water multiply and intensify, tension between Hindus and Muslims could ignite, leading to instability.

Fortunately, state failure is not always a one-way street. South Africa, which could have erupted into a race war a generation ago, is now a functioning democracy. Liberia and Colombia, both of which once had high Failed State Index scores, have each made a remarkable turnaround.

Nevertheless, as the number of failing states grows, dealing with various international crises becomes more difficult. Situations that may be manageable in a healthy world order, such as maintaining monetary stability or controlling an infectious disease outbreak, become difficult and sometimes impossible in a world with many disintegrating states. Even maintaining international flows of raw materials could become a challenge. At some point, spreading political instability could disrupt global economic progress, underscoring the need to address the causes of state failure with a heightened sense of urgency.

III
THE RESPONSE: PLAN B

Many earlier civilizations faced environmentally induced crises and many were undone by them. Typically, they faced one or two destructive environmental trends, most often deforestation and soil erosion. In contrast, our early twenty-first century global civilization faces numerous environmentally damaging trends, all of our own making and many of them reinforcing each other. In addition to deforestation and soil erosion, they include aquifer depletion, crop-withering heat waves, collapsing fisheries, melting mountain glaciers, and rising sea level, to cite a few.

While the preceding seven chapters described these trends and their consequences, notably environmental refugees and failing states, the next five chapters describe what it will take to reverse these trends.

Because the world today is ecologically and economically interdependent, today's environmental crises are uniquely global in scope. In this new world, the term national security has little meaning because we will either all make it together or go down together.

Simply put, what the Earth Policy Institute calls Plan B is what we have to do to save civilization. It is a monumental effort to be undertaken at wartime speed. There is no historical precedent simply because the entire world has never before been so threatened.

As noted in Chapter 1, Plan B has four components: stabilizing climate, restoring the earth's natural support systems, stabilizing population, and eradicating poverty. The climate stabilization plan calls for an 80 percent reduction in carbon dioxide (CO_2) emissions by 2020. In establishing this goal, we did not ask what might be politically popular but rather what was needed if we want to have any hope of saving the Greenland ice sheet and at least the larger glaciers in the Himalayas and on the Tibetan plateau.

Reducing carbon emissions has three main elements. The first is raising the efficiency of the world energy economy while restructuring the transport sector. This is designed to offset all projected growth in energy use between now and 2020. The second is cutting emissions in the energy sector, principally by replacing fossil fuels (oil, coal, and natural gas) with renewable energy (wind, solar, and geothermal). The third element in cutting carbon emissions is to end deforestation while engaging in a massive campaign to plant trees and stabilize soils.

The 80 percent reduction in CO_2 emissions is designed to bring the rise in atmospheric CO_2 concentrations, currently at 387 parts per million (ppm), to an end by 2020 at 400 ppm. Once we stop the rise, then we can begin to reduce CO_2 concentrations to the 350 ppm that leading climate scientists recommend.

The other three Plan B components go hand in hand. The restoration of the earth's natural systems—including reforestation, soil conservation, fishery restoration, and aquifer stabilization—will help us eradicate poverty. Likewise, eradicating poverty helps stabilize population, and accelerating the shift to smaller families helps people break out of poverty. Ultimately, feeding a global population of 8 billion will depend on our success in meeting all four Plan B goals.

The good news is that we have the resources needed to achieve this. The restructuring of the energy economy, including both the shift to more energy-efficient technologies and the replacement of fossil fuels with renewable sources, can be achieved largely by lowering the tax on incomes and raising the tax on carbon. Plan B calls for phasing in a worldwide carbon tax of $200 per ton by 2020, while offsetting it at each step of the way with a reduction in income taxes.

The budget for restoring the earth's natural systems, stabilizing population, and eradicating poverty will require under $200 billion per year in additional expenditures. This can be achieved simply by updating the concept of national security to recognize the new threats to our security and reallocating the security budget accordingly.

The final chapter in this book is about mobilization. It talks about the various models of social change, how to achieve a rapid transformation of society, and the urgency of implementing Plan B.

8

Building an Energy-Efficient Global Economy

Advancing technologies offer the world a greater potential for cutting energy use today than at any time in history. For example, during much of the last century nearly all the household light bulbs on the market were inefficient incandescents. But today people can also buy compact fluorescent lamps (CFLs) that use only one fourth as much electricity. And the light-emitting diodes (LEDs) now coming to market use even less.

A similar situation exists with cars. During the century since the automobile appeared, an internal combustion engine was the only option. Now we can buy plug-in hybrids and all-electric cars that run largely or entirely on electricity. And since an electric motor is over three times as efficient as an internal combustion engine, there is an unprecedented potential for reducing energy use in the transport sector.

Beyond energy-saving technologies, vast amounts of energy can be saved by restructuring key sectors of the economy. Designing cities for people, not for cars, is a great place to begin. And if we can move beyond the throwaway society, reusing and recycling almost everything, imagine how much material and energy we can save.

One of the quickest ways to cut carbon emissions and save money is simply to change light bulbs. Replacing inefficient incandescent bulbs with CFLs can reduce the electricity used for lighting by three fourths. And since they last up to 10 times as long, each standard CFL will cut electricity bills by roughly $40 over its lifetime.

The world has reached a tipping point in shifting to compact fluorescents, as many countries phase out incandescents. But even before the transition is complete, the shift to LEDs is under way. Now the world's most advanced lighting technology, the LED uses even less energy than a CFL and up to 85 percent less than an incandescent. And LEDs offer another strong economic advantage—longevity. An LED installed when a child is born is likely to still be working when the youngster graduates from college.

With costs falling fast, LEDs are quickly taking over several niche markets, such as traffic lights. In the United States, almost 70 percent of traffic lights have been converted to LEDs, while the figure is still less than 20 percent in Europe. New York City has changed all its traffic lights to LEDs, cutting the annual bill for power and maintenance by $6 million.

For the far more numerous street lights, the potential savings are even greater. In 2009, Los Angeles Mayor Antonio Villaraigosa said the city would replace its 140,000 street lights with LEDs, saving taxpayers $48 million over seven years. With replacement well along, the electricity bill for street lights was down 55 percent as of mid-2010.

Leading bulb manufacturers such as Phillips and GE are currently selling their lower-wattage LEDs for $20. As prices fall, Zia Eftekhar, head of Phillips lighting in North America, expects LEDs to take more than 50 percent of the North American and European markets by 2015 and 80 percent by 2020. In 2009, China and Taiwan joined forces in manufacturing LEDs to compete more effectively with Japan (currently the world leader), South Korea, Germany, and the United States.

Energy can also be saved by using motion sensors that turn lights off in unoccupied spaces. Automatic dimmers can reduce the intensity of interior lighting when sunlight is bright. In fact, LEDs combined with these "smart" lighting technologies can cut electricity bills by 90 percent compared with incandescents.

All told, shifting to CFLs in homes, to the most advanced linear fluorescents in office buildings, commercial outlets, and factories, and to LEDs for traffic lights would cut the world share of electricity used for lighting from 19 to 7 percent. This would save enough electricity to close 705 of the world's 2,800 coal-fired plants. If the world turns heavily to LEDs for lighting by 2020, as now seems likely, the savings would be even greater.

A similar range of efficiencies is available for many household appliances. Although the U.S. Congress has been passing legislation since 1975 to raise efficiency for 22 broad categories of household and industrial appliances, from dishwashers to electric motors, the U.S. Department of Energy (DOE) had failed to write the standards needed to implement the legislation. To remedy this, just days after taking office President Barack Obama ordered DOE to write the needed regulations and thus tap this reservoir of efficiency. In September 2010, DOE announced that new efficiency standards for more than 20 household and commercial products had been finalized since January 2009, noting that this "will cumulatively save consumers between $250 billion and $300 billion through 2030."

A more recent efficiency challenge is presented by large flat-screen televisions. The screens now on the market use much more electricity than traditional cathode ray tube televisions—indeed, nearly four times as much if they are large-screen plasma models. Setting the U.S. pace in this area, as in so many others, California is requiring that all new televisions draw one third less electricity than current sets do by 2011 and 49 percent less by 2013. Because the California market is so large, it could very likely force the industry to meet this standard nationwide.

The big appliance efficiency challenge is China, where modern appliance ownership in cities today is similar to that in industrial countries. For every 100 urban households there are 133 color TV sets, 95 washing machines, and 100 room air conditioners. This phenomenal growth, with little attention to efficiency, helped raise China's electricity use a staggering 11-fold from 1980 to 2007.

Along with the United States and China, the European Union has the other major concentration of home appliances. Greenpeace notes that even though Europeans on average use half as much electricity as Americans do, they still have a large reduction potential. A refrigerator in Europe uses scarcely half as much electricity as one in the United States, for example, but the most efficient refrigerators on the market today use only one fourth as much electricity as the average refrigerator in Europe, suggesting a huge potential for cutting electricity use further everywhere.

Technological progress keeps raising the potential for efficiency gains. Japan's Top Runner Program is the world's most dynamic system for upgrading appliance efficiency standards. In this system, the most efficient appliances marketed today set the standard for those sold tomorrow. Within a decade, Japan raised efficiency standards for individual appliances by anywhere from 15 to 83 percent. This ongoing process continually exploits advances in efficiency technologies.

Although appliances account for a significant share of electricity use in buildings, heating and cooling require more energy in total. But buildings often get short shrift in efficiency planning, even though the sector is the leading source of carbon emissions, eclipsing transportation.

Because buildings last for 50–100 years or longer, it is often assumed that cutting carbon emissions in this sector is a long-term process. But that is not necessarily the case. An energy retrofit of an older inefficient building can cut energy bills by 20–50 percent or more. The next step, shifting entirely to renewable sources of electricity to heat, cool, and light the building, completes the job. Presto! A zero-carbon building.

In the United States, the stimulus package signed by President Obama in February 2009 provided for weatherizing a million private homes, weatherizing and retrofitting part of the nation's public housing, and making government buildings more energy-efficient. These initiatives are intended to help build a vigorous U.S. energy efficiency industry.

Among the numerous efforts to make older structures more efficient is the Clinton Foundation's Energy Efficiency Building Retrofit Program, a project of the Clinton Climate Initiative. In cooperation with C40, a large-cities climate leadership group, this program brings together financial institutions and some of the world's largest energy service and technology companies to work with cities to retrofit buildings, reducing their energy use by up to 50 percent. The energy service companies—including Johnson Controls and Honeywell—committed to provide building owners with contractual "performance guarantees" assuring the energy savings and maximum costs of the retrofit project. At the launch of this program, former President Bill Clinton pointed out that banks and energy service companies would make money, building owners would save money, and carbon emissions would fall.

In April 2009, the owners of New York's Empire State Building announced plans to retrofit the iconic 80-year-old 102-story building, reducing its energy use by nearly 40 percent. The resulting annual energy savings of $4.4

million is expected to recover the retrofitting costs in three years.

The carbon reductions from retrofitting are impressive, but new buildings can be designed to emit far less carbon. As of January 2009, Germany required that all new buildings either get at least 15 percent of water and space heating from renewable energy or dramatically improve the efficiency with which they use energy. The bonus here is that if a builder is putting a solar water and space heater on the roof, it is unlikely that it will be limited to meeting only 15 percent of the building's needs.

One firm believer in the potential for cutting energy use in new buildings is Edward Mazria, a climate-conscious architect from New Mexico who has launched the 2030 Challenge. Its principal goal is to get U.S. architects to design all buildings by 2030 to operate without fossil fuels. Mazria notes that "it's the architects who hold the key to turning down the global thermostat." To reach his goal, Mazria has organized a coalition of several organizations, including the American Institute of Architects, the U.S. Green Building Council (USGBC), and the U.S. Conference of Mayors.

In the private sector, the USGBC—well known for its Leadership in Energy and Environmental Design (LEED) certification and rating program—heads the field. This voluntary program has four certification levels—certified, silver, gold, and platinum. A LEED-certified building must meet minimal standards in environmental quality, materials use, energy efficiency, water efficiency, and site selection, which includes access to public transit. LEED-certified buildings are attractive to buyers because they have lower operating costs, higher lease rates, and happier, healthier occupants than traditional buildings do.

The Chesapeake Bay Foundation's office building for its 100 staff members near Annapolis, Maryland, was the first to earn a LEED platinum rating. Among its features are a ground-source heat pump for heating and cooling, a rooftop solar water heater, and sleekly designed composting toilets that produce a rich humus used to fertilize the landscape surrounding the building.

A 60-story office building with a gold rating built in Chicago uses river water to cool the building in summer and has covered over half the rooftop with plants to reduce runoff and heat loss. The principal tenant, Kirkland and Ellis LLP, a Chicago-based law firm, insisted that the building be at least silver-certified and that this be incorporated into the lease.

The 55-story Bank of America tower in New York is the first large skyscraper to have earned a platinum rating. It has its own co-generation power plant and collects rainwater, reuses waste water, and used recycled materials in construction. Worldwide, Pike Research projects the floor area of buildings certified by green building standards to expand from 6 billion square feet in 2010 to 53 billion feet by 2020.

Within the transportation sector itself, there are numerous opportunities for energy savings. The first step in increasing efficiency and cutting carbon emissions is to simultaneously restructure and electrify the transport system to facilitate the shift from fossil fuels to renewable electricity. Restructuring involves strengthening urban public transportation and designing communities to reduce the need for cars. For traveling between cities, developing a high-speed intercity rail system, similar to those in Japan, Western Europe, and China, is the key.

Urban transport systems based on a combination of subways, light rail, bus lines, bicycle pathways, and pedestrian walkways offer the best of all possible worlds in providing mobility, low-cost transportation, and a healthy urban environment. And since rail systems are geographically fixed, the nodes on such a system become the obvious places to concentrate high-rise office and apartment buildings as well as shops.

Some of the most innovative public transportation systems have evolved in developing-country cities such as Bogotá, Colombia. The success of Bogotá's bus rapid transit (BRT) system, which uses special express lanes to move people quickly through the city, is being replicated in scores of other cities, including Mexico City, São Paulo, Hanoi, Seoul, Istanbul, and Quito. In China, BRT systems operate in 11 cities, including Beijing.

In Paris, Mayor Bertrand Delanoë inherited some of Europe's worst traffic congestion and air pollution when he was elected in 2001. The first of three steps he took to reduce traffic was to invest in more-accessible high-quality public transit throughout the greater Paris area. The next step was to create express lanes on main thoroughfares for buses and bicycles, thus reducing the number of lanes for cars. As the speed of buses increased, more people used them.

A third innovative initiative in Paris was the establishment of a city bicycle rental program that has 24,000 bikes available at 1,750 docking stations throughout the city. Rates for rental range from just over $1 per day to $40 per year, but if the bike is used for fewer than 30 minutes, the ride is free. Based on the first two years, the bicycles are proving to be immensely popular—with 63 million trips taken as of late 2009. Hundreds of other cities, including London, Washington, Shanghai, Mexico City, and Santiago are also adopting urban bicycle rental systems. Bicycle sharing is an idea whose time has come.

Any serious global effort to cut automotive fuel use begins with the United States, which consumes more gasoline than the next 20 countries combined, including Japan, China, Russia, Germany, and Brazil. The United States—with 248 million passenger vehicles out of the global 965 million—not only has by far the largest fleet of cars but is near the top in miles driven per car and near the bottom in vehicle fuel efficiency.

The car promised mobility, and in a largely rural society it delivered. But the growth in urban car numbers at some point provides not mobility, but immobility. The Texas Transportation Institute reports that U.S. congestion costs, including fuel wasted and time lost, climbed from $17 billion in 1982 to $87 billion in 2007.

Many American communities lack sidewalks and bike lanes, making it difficult for pedestrians and cyclists to get around safely, particularly where streets are heavily traveled. Fortunately, the country that has lagged far behind Europe in developing diversified urban transport systems is being swept by a "complete streets" movement, an effort to ensure that streets are friendly to pedestrians and bicycles as well as to cars.

The National Complete Streets Coalition, a powerful assemblage of citizen groups, including the Natural Resources Defense Council, AARP (an organization of nearly 40 million older Americans), and numerous local and national cycling organizations is challenging the "cars only" model. As of October 2010, complete streets policies were in place in 23 states, including more populous states like California and Illinois, and in 98 cities.

America's century-old love affair with the automobile may be coming to an end. The U.S. fleet has apparently peaked. In 2009, the 12.4 million cars scrapped exceeded the 10.6 million new cars sold, shrinking the fleet by nearly 1 percent. While this has been widely associated with the recession, it was in fact caused by several converging forces, including market saturation, ongoing urbanization, economic uncertainty, oil insecurity, rising gasoline prices, frustration with traffic congestion, and mounting concerns about climate change.

Perhaps the leading social trend affecting the future of the automobile is the declining interest in cars among young people. For past generations, growing up in a country that was still heavily rural, getting a driver's license and a car or a pickup was a rite of passage. In contrast, now that the United States is 82 percent urban, more young Americans are growing up in families without cars. They socialize on the Internet and on smartphones, not in cars. Many do not even bother to get a driver's license. Because of these converging trends, I believe that the U.S. fleet could shrink 10 percent by 2020. Japan's fleet, second in size to the U.S. fleet, is also shrinking.

Beyond shrinking the fleet, the key to reducing U.S. gasoline use in the near term is to raise fuel efficiency standards. The 40-percent increase in the fuel efficiency of new cars by 2016 announced by the Obama administration in May 2009 will reduce both carbon emissions and dependence on oil. A crash program to shift the U.S. fleet to plug-in hybrids and all-electric cars could make an even greater contribution. And shifting public funds from highway construction to public transit and intercity rail would further reduce the number of cars needed, bringing the United States closer to the Plan B goal of cutting carbon emissions 80 percent by 2020.

Plug-in hybrids and all-electric cars are coming to market. The Chevrolet Volt plug-in hybrid is scheduled to be available in late 2010. At the same time, Nissan will be bringing its all-electric car, the Leaf, to market in the United States, Japan, and Europe. And in 2012, Toyota plans to release a plug-in version of its popular Prius hybrid. With the transition to renewable energy gaining momentum, cars could one day run largely on wind-generated electricity that costs the equivalent of less than $1 per gallon of gasoline.

Shifting to plug-in electric hybrids and all-electric cars does not require a costly new infrastructure, since the network of gasoline service stations and the electricity grid are already in place. A 2006 study by the U.S. Pacific

Northwest National Laboratory estimated that over 70 percent of the electricity needs of a national fleet of plug-in cars could be satisfied with the existing electricity supply, since the recharging would take place largely at night when there is an excess of generating capacity. What will be needed in addition to home hookups are readily accessible electrical outlets in parking garages, parking lots, and street-side parking meters to facilitate recharging.

Few methods of reducing carbon emissions are as effective as substituting a bicycle for a car on short trips. A bicycle is a marvel of engineering efficiency, one where an investment in 22 pounds of metal and rubber boosts personal mobility by a factor of three. On my bike I estimate that I get easily 7 miles per potato. An automobile, which typically requires at least a ton of material to transport one person, is extraordinarily inefficient by comparison.

The bicycle has many attractions as a form of personal transportation. It is carbon-free, alleviates congestion, lowers air pollution, reduces obesity, and is priced within the reach of billions of people who cannot afford a car. Bicycles increase mobility while reducing congestion and the area of land paved over. As bicycles replace cars, cities can convert parking lots into parks or urban gardens.

As campuses are overwhelmed by cars, and with the construction of parking garages costing $55,000 per parking space, colleges, like cities, are turning to bikes. Chicago's St. Xavier University launched a bike-sharing program in the fall of 2008, with students using their ID cards instead of credit cards. Emory University in Atlanta, Georgia, has introduced a free bike-sharing system. Ripon College in Wisconsin and the University of New England in Maine have gone even further: they give a bike to freshmen who agree to leave their cars at home.

The key to realizing the bicycle's potential is to create a bike-friendly transport system. This means providing both bicycle trails and designated street lanes for bicycles and then linking them with public transit options. Among the industrial-country leaders in designing bicycle-friendly transport systems are the Netherlands, where 25 percent of all trips are by bike, Denmark with 18 percent, and Germany, 10 percent. For the United States, the equivalent figure is 1 percent.

While the future of transportation in cities lies with a mix of light rail, buses, bicycles, cars, and walking, the future of intercity travel belongs to high-speed trains. Japan's bullet trains, operating at up to 190 miles per hour, carry nearly 400,000 passengers a day. On some heavily used intercity lines, trains depart every three minutes.

Over the last 46 years, Japan's high-speed trains have carried billions of passengers in great comfort without a fatal crash. Late arrivals average 6 seconds. If we were selecting seven wonders of the modern world, Japan's high-speed rail system surely would be among them.

Although the first European high-speed line, from Paris to Lyon, did not begin operation until 1981, Europe has since made enormous strides. As of 2010 there were 3,800 miles of high-speed rail operating in Europe. The goal is to triple this track length by 2025 and eventually to integrate the East European countries into a continental network.

High-speed intercity rail links are changing travel patterns by reducing long drives and short flights, each of which is carbon-intensive. When the Brussels-to-Paris link opened, the share of people traveling between the two cities by train rose from 24 to 50 percent. The car share dropped from 61 to 43 percent, and plane travel virtually disappeared.

While France and Germany were the early European leaders in intercity rail, Spain is quickly building a high-speed rail network that is enormously popular. Before the recent Barcelona-to-Madrid high-speed rail connection, 90 percent of the 6 million trips between the two cities each year were by air. By early 2010, more people were making the trip by train than by plane. By 2020, half of the country's transportation budget will be going to rail. As *The Economist* notes, "Europe is in the grip of a high-speed rail revolution."

Until recently, there was a huge gap in high-speed rail between Japan and Europe, on the one hand, and the rest of the world on the other. That is changing as China moves to the fore with both the world's fastest trains and the most ambitious high-speed rail construction program of any country. For various reasons, including land scarcity and oil dependency, China is shifting the emphasis from building American-style expressways to building an intercity network of high-speed trains linked directly to urban subway systems, some 60 of which are under construction. The goal is to reduce the need for cars and planes for medium and longer distance travel. When a 300-mile-long line opened in 2010 between Zhengzhou and Xi'an, the low-cost, two-hour train ride was so popular that all flights between the two cities were discontinued.

China is spending $120 billion on high-speed rail in 2010, whereas the United States is spending $1 billion. While the United States allocated $8 billion for high-speed rail from its stimulus package, China allocated $100 billion of its stimulus funding to this cause. It thus comes as no surprise that by 2012 China will have more high-speed rail track mileage than the rest of the world combined.

The United States has a "high-speed" Acela Express that links Washington, New York, and Boston, but unfortunately neither its average speed of 70 miles per hour nor its reliability remotely approach those of the trains in Japan, Europe, and now China.

It is time for the United States to shift investment from

roads and highways to railways to build a twenty-first century transport system. In 1956, President Eisenhower launched the interstate highway system, justifying it on national security grounds. Today, both climate change and oil insecurity argue for the construction of a national high-speed rail system.

Carbon dioxide emissions per passenger mile on high-speed trains are roughly one third those of cars and one fourth those of planes. In the Plan B economy, carbon emissions from trains will essentially be zero, since they will be powered by wind, solar, and geothermal electricity. In addition to being comfortable and convenient, these rail links reduce air pollution and congestion.

Restructuring the transportation system also has a huge potential for reducing materials use as light rail and buses replace cars. For example, 60 cars, weighing a total of 110 tons, can be replaced by one 12-ton bus, reducing material use 89 percent.

Savings from replacing a car with a bike are even more impressive. Urban planner Richard Register recounts meeting a bicycle-activist friend wearing a T-shirt that said, "I just lost 3,500 pounds. Ask me how." When queried, he said he had sold his car. Replacing a 3,500-pound car with a 22-pound bicycle obviously reduces fuel use dramatically, but it also reduces materials use by 99 percent, indirectly saving still more energy.

The production, processing, and disposal of materials in our modern throwaway economy wastes not only materials but the energy embodied in the material as well. The throwaway economy that has evolved over the last half-century is an aberration that is now itself headed for the junk heap of history.

In their book *Cradle to Cradle: Remaking the Way We Make Things*, American architect William McDonough and German chemist Michael Braungart conclude that waste and pollution are to be avoided entirely. "Pollution," says McDonough, "is a symbol of design failure."

Cutting the use of virgin raw materials begins with recycling steel, the use of which dwarfs that of all other metals combined. In the United States, virtually all cars are recycled. They are simply too valuable to be left to rust in out-of-the-way junkyards. With the number of cars scrapped now exceeding new cars sold, the U.S. automobile sector actually has a steel surplus that can be used elsewhere in the economy. The U.S. recycling rate for household appliances is estimated at 90 percent. For steel cans it is 65 percent. For construction steel, the figures are 98 percent for steel beams and girders but only 65 percent for reinforcement steel.

Beyond reducing materials use, the energy savings from recycling are huge. Making steel from recycled scrap takes only 26 percent as much energy as that from iron ore. For aluminum, the figure is just 4 percent. Recycled plastic uses only 20 percent as much energy. Recycled paper uses 64 percent as much—and with far fewer chemicals during processing. If the world recycling rates of these basic materials were raised to those already attained in the most efficient economies, world carbon emissions would drop precipitously.

In the United States, only 33 percent of garbage is recycled. Some 13 percent is burned and 54 percent goes to landfills, indicating a huge potential for reducing materials use, energy use, and pollution. Among the larger U.S. cities, recycling rates vary from 25 percent in New York to 45 percent in Chicago, 65 percent in Los Angeles, and 77 percent in San Francisco, the highest of all.

One way to encourage recycling is simply to adopt a landfill tax. For example, when the small town of Lyme, New Hampshire, adopted a pay-as-you-throw (PAYT) program that encourages municipalities to charge residents for each bag of garbage, it dramatically reduced the flow of materials to landfills, raising the share of garbage recycled from 13 to 52 percent in only one year, simultaneously reducing the town's landfill fees, and generating a cash flow from the sale of recycled material. Nationwide, more than 7,000 U.S. communities now have PAYT programs.

In addition to measures that encourage recycling, there are those that encourage or mandate the reuse of products such as refillable beverage containers. Finland, for example, has banned the use of one-way soft drink containers. A refillable glass bottle used over and over requires only 10 percent as much energy per use as recycling an aluminum can. Banning nonrefillables is a quintuple win option—cutting material use, carbon emissions, air pollution, water pollution, and landfill costs simultaneously.

Bottled water is even more wasteful. In a world trying to stabilize climate, it is difficult to justify bottling water (often tap water to begin with), hauling it long distances, and then selling it for 1,000 times the price of water from the kitchen faucet. Although clever marketing has convinced many consumers that bottled water is safer and healthier than tap water, a detailed study by WWF found that in the United States and Europe there are more standards regulating the quality of tap water than there are for bottled water. In developing countries where water is unsafe, it is far cheaper to boil or filter water than to buy it in bottles.

Manufacturing the nearly 28 billion plastic bottles used each year to package water in the United States alone requires the equivalent of 17 million barrels of oil. This—combined with the energy used to refrigerate and haul the bottled water in trucks, sometimes over hundreds of miles—means the U.S. bottled water industry consumes roughly 50 million barrels of oil per year, equal to 13 percent of U.S. oil imports from Saudi Arabia.

The potential for reducing energy use across the

board is huge. For the United States, the Rocky Mountain Institute calculates that if the 40 least efficient states were to achieve the electrical efficiency of the 10 most efficient ones, national electricity use would be cut by one third. This alone would allow the equivalent of 62 percent of all U.S. coal-fired power plants to be closed down. But even the most efficient states have a substantial potential for reducing electricity use further and, indeed, are planning to keep cutting carbon emissions and saving money.

The opportunities to save energy are everywhere, permeating every corner of the economy, every facet of our lives, and every country. Exploiting this abundance of wasted energy will allow the world to actually reduce total energy use over the next decade. These potentially massive efficiency gains, combined with the worldwide shift to renewable energy outlined in the next chapter, will move the world ever closer to the Plan B energy economy.

Data, endnotes, and additional resources can be found on Earth Policy's Web site, at www.earth-policy.org.

9

Harnessing Wind, Solar, and Geothermal Energy

As fossil fuel prices rise, as oil insecurity deepens, and as concerns about climate change cast a shadow over the future of coal, a new world energy economy is emerging. The old energy economy, fueled by oil, coal, and natural gas, is being replaced with an economy powered by wind, solar, and geothermal energy. Despite the global economic crisis, this energy transition is moving at a pace and on a scale that we could not have imagined even two years ago.

The transition is well under way in the United States, where both oil and coal consumption have recently peaked. Oil consumption fell 8 percent between 2007 and 2010 and will likely continue falling over the longer term. During the same period, coal use also dropped 8 percent as a powerful grassroots anti-coal movement brought the licensing of new coal plants to a near standstill and began to work on closing existing ones.

While U.S. coal use was falling, some 300 wind farms with a generating capacity of 21,000 megawatts came online. Geothermal generating capacity, which had been stagnant for 20 years, came alive. In mid-2010, the U.S.-based Geothermal Energy Association announced that 152 new geothermal power plants were being developed, enough to triple U.S. geothermal generating capacity. On the solar front, solar cell installations are doubling every two years. The dozens of U.S. solar thermal power plants in the works could collectively add some 9,900 megawatts of generating capacity.

This chapter lays out the worldwide Plan B goals for developing renewable sources of energy by 2020. The goal of cutting carbon emissions 80 percent by 2020 is based on what we think is needed to avoid civilization-threatening climate change. This is not Plan A, business as usual. This is Plan B—a wartime mobilization, an all-out effort to restructure the world energy economy.

To reach the Plan B goal, we replace all coal- and oil-fired electricity generation with that from renewable sources. Whereas the twentieth century was marked by the globalization of the world energy economy as countries everywhere turned to oil, much of it coming from the Middle East, this century will see the localization of energy production as the world turns to wind, solar, and geothermal energy.

The Plan B energy economy, which will be powered largely by electricity, does not rely on a buildup in nuclear power. If we used full-cost pricing—insisting that utilities pay for disposing of nuclear waste, decommissioning worn-out plants, and insuring reactors against possible accidents and terrorist attacks—no one would build a nuclear plant. They are simply not economical. Plan B also excludes the oft-discussed option of capturing and sequestering carbon dioxide (CO_2) from coal-fired power plants. Given the costs and the lack of investor interest within the coal community itself, this technology is not likely to be economically viable by 2020, if ever.

Instead, wind is the centerpiece of the Plan B energy economy. It is abundant, low cost, and widely distributed; it scales up easily and can be developed quickly. A 2009 survey of world wind resources published by the U.S. National Academy of Sciences reports a wind-generating potential on land that is 40 times the current world consumption of electricity from all sources.

For many years, a small handful of countries dominated growth in wind power, but this is changing as the industry goes global, with more than 70 countries now developing wind resources. Between 2000 and 2010, world wind electric generating capacity increased at a frenetic pace from 17,000 megawatts to nearly 200,000 megawatts.

The United States, with 35,000 megawatts of wind generating capacity, leads the world in harnessing wind, followed by China and Germany with 26,000 megawatts each. Texas, long the leading U.S. oil-producing state, is now also the nation's leading generator of electricity from wind. It has 9,700 megawatts of wind generating capacity online, 370 megawatts more under construction, and a huge amount under development. If all of the wind farms projected for 2025 are completed, Texas will have 38,000 megawatts of wind generating capacity—the equivalent of 38 coal-fired power plants. This would satisfy roughly 90 percent of the current residential electricity needs of the state's 25 million people.

In July 2010, ground was broken for the Alta Wind Energy Center (AWEC) in the Tehachapi Pass, some 75 miles north of Los Angeles, California. At 1,550 megawatts, it will be the largest U.S. wind farm. The AWEC is part of what will eventually be 4,500 megawatts of renewable power generation, enough to supply electricity to some 3 million homes.

Since wind turbines occupy only 1 percent of the land covered by a wind farm, farmers and ranchers can continue to grow grain and graze cattle on land devoted to wind farms. In effect, they double-crop their land, simultaneously harvesting electricity and wheat, corn, or cattle. With no investment on their part, farmers and ranchers typically receive $3,000–10,000 a year in royalties for each wind turbine on their land. For thousands of ranchers in the U.S. Great Plains, wind royalties will dwarf their net earnings from cattle sales.

In considering the energy productivity of land, wind turbines are in a class by themselves. For example, an acre of land in northern Iowa planted in corn can yield $1,000 worth of ethanol per year. That same acre used to site a wind turbine can produce $300,000 worth of electricity per year. This helps explain why investors find wind farms so attractive.

Impressive though U.S. wind energy growth is, the expansion now under way in China is even more so. China has enough onshore harnessable wind energy to raise its current electricity consumption 16-fold. Today, most of China's 26,000 megawatts of wind generating capacity come from 50- to 100-megawatt wind farms. Beyond the many other wind farms of that size that are on the way, China's new Wind Base program is creating seven wind mega-complexes of 10 to 38 gigawatts each in six provinces (1 gigawatt equals 1,000 megawatts). When completed, these complexes will have a generating capacity of more than 130 gigawatts. This is equivalent to building one new coal plant per week for two and a half years.

Of these 130 gigawatts, 7 gigawatts will be in the coastal waters of Jiangsu Province, one of China's most highly industrialized provinces. China is planning a total of 23 gigawatts of offshore wind generating capacity. The country's first major offshore project, the 102-megawatt Donghai Bridge Wind Farm near Shanghai, is already in operation.

In Europe, which now has 2,400 megawatts of offshore wind online, wind developers are planning 140 gigawatts of offshore wind generating capacity, mostly in the North Sea. There is enough harnessable wind energy in offshore Europe to satisfy the continent's needs seven times over.

In September 2010, the Scottish government announced that it was replacing its goal of 50 percent renewable electricity by 2020 with a new goal of 80 percent. By 2025, Scotland expects renewables to meet all of its electricity needs. Much of the new capacity will be provided by offshore wind.

Measured by share of electricity supplied by wind, Denmark is the national leader at 21 percent. Three north German states now get 40 percent or more of their electricity from wind. For Germany as a whole, the figure is 8 percent—and climbing. And in the state of Iowa, enough wind turbines came online in the last few years to produce up to 20 percent of that state's electricity.

Denmark is looking to push the wind share of its electricity to 50 percent by 2025, with most of the additional power coming from offshore. In contemplating this prospect, Danish planners have turned conventional energy policy upside down. They plan to use wind as the mainstay of their electrical generating system and to use fossil-fuel-generated power to fill in when the wind dies down.

Spain, which has 19,000 megawatts of wind-generating capacity for its 45 million people, got 14 percent of its electricity from wind in 2009. On November 8th of that year, strong winds across Spain enabled wind turbines to supply 53 percent of the country's electricity over a five-hour stretch. London *Times* reporter Graham Keeley wrote from Barcelona that "the towering white wind turbines which loom over Castilla-La Mancha—home of Cervantes's hero, Don Quixote—and which dominate other parts of Spain, set a new record in wind energy production."

In 2007, when Turkey issued a request for proposals to build wind farms, it received bids to build a staggering 78,000 megawatts of wind generating capacity, far beyond its 41,000 megawatts of total electrical generating capacity. Having selected 7,000 megawatts of the most promising proposals, the government is issuing construction permits.

In wind-rich Canada, Ontario, Quebec, and Alberta are the leaders in installed capacity. Ontario, Canada's most populous province, has received applications for offshore wind development rights on its side of the Great Lakes that could result in some 21,000 megawatts of generating capacity. The provincial goal is to back out all coal-fired power by 2014.

On the U.S. side of Lake Ontario, New York State is also requesting proposals. Several of the seven other states that border the Great Lakes are planning to harness lake winds.

At the heart of Plan B is a crash program to develop 4,000 gigawatts (4 million megawatts) of wind generating capacity by 2020, enough to cover over half of world electricity consumption in the Plan B economy. This will require a near doubling of capacity every two years, up from a doubling every three years over the last decade.

This climate-stabilizing initiative would mean the installation of 2 million wind turbines of 2 megawatts each. Manufacturing 2 million wind turbines over the next 10 years sounds intimidating—until it is compared with the 70 million automobiles the world produces each year.

At $3 million per installed turbine, the 2 million turbines would mean spending $600 billion per year worldwide between now and 2020. This compares with world oil and gas capital expenditures that are projected to double from $800 billion in 2010 to $1.6 trillion in 2015.

The second key component of the Plan B energy economy is solar energy, which is even more ubiquitous than wind energy. It can be harnessed with both solar photovoltaics (PV) and solar thermal collectors. Solar PV—both silicon-based and thin film—converts sunlight directly into electricity. A large-scale solar thermal technology, often referred to as concentrating solar power (CSP), uses reflectors to concentrate sunlight on a liquid, producing steam to drive a turbine and generate electricity. On a smaller scale, solar thermal collectors can capture the sun's radiant energy to warm water, as in rooftop solar water heaters.

The growth in solar cell production can only be described as explosive. It climbed from an annual expansion of 38 percent in 2006 to an off-the-chart 89 percent in 2008, before settling back to 51 percent in 2009. At the end of 2009, there were 23,000 megawatts of PV installations worldwide, which when operating at peak power could match the output of 23 nuclear power plants.

On the manufacturing front, the early leaders—the United States, Japan, and Germany—have been overtaken by China, which produces more than twice as many solar cells annually as Japan. Number three, Taiwan, is moving fast and may overtake Japan in 2010. World PV production has roughly doubled every two years since 2001 and will likely approach 20,000 megawatts in 2010.

Germany, with an installed PV power generating capacity of almost 10,000 megawatts, is far and away the world leader in installations. Spain is second with 3,400 megawatts, followed by Japan, the United States, and Italy. Ironically, China, the world's largest manufacturer of solar cells, has an installed capacity of only 305 megawatts, but this is likely to change quickly as PV costs fall.

Historically, photovoltaic installations were small-scale—mostly residential rooftop installations. Now that is changing as utility-scale PV projects are being launched in several countries. The United States, for example, has under construction and development some 77 utility-scale projects, adding up to 13,200 megawatts of generating capacity. Morocco is now planning five large solar-generating projects, either photovoltaic or solar thermal or both, each ranging from 100 to 500 megawatts in size.

More and more countries, states, and provinces are setting solar installation goals. Italy's solar industry group is projecting 15,000 megawatts of installed capacity by 2020. Japan is planning 28,000 megawatts by 2020. The state of California has set a goal of 3,000 megawatts by 2017.

Solar-rich Saudi Arabia recently announced that it plans to shift from oil to solar energy to power new desalination plants that supply the country's residential water. It currently uses 1.5 million barrels of oil per day to operate some 30 desalting plants.

With installations of solar PV climbing, with costs continuing to fall, and with concerns about climate change escalating, cumulative PV installations could reach 1.5 million megawatts (1,500 gigawatts) in 2020. Although this estimate may seem overly ambitious, it could in fact be conservative, because if most of the 1.5 billion people who lack electricity today get it by 2020, it will likely be because they have installed home solar systems. In many cases, it is cheaper to install solar cells for individual homes than it is to build a grid and a central power plant.

The second, very promising way to harness solar energy on a massive scale is CSP, which first came on the scene with the construction of a 350-megawatt solar thermal power plant complex in California. Completed in 1991, it was the world's only utility-scale solar thermal generating facility until the completion of a 64-megawatt power plant in Nevada in 2007.

Two years later, in July 2009, a group of 11 leading European firms and one Algerian firm, led by Munich Re and including Deutsche Bank, Siemens, and ABB, announced that they were going to craft a strategy and funding proposal to develop solar thermal generating capacity in North Africa and the Middle East. Their proposal would meet the needs of the producer countries and supply part of Europe's electricity via undersea cable.

This initiative, known as the Desertec Industrial Initiative, could develop 300,000 megawatts of solar thermal generating capacity—huge by any standard. It is driven by concerns about disruptive climate change and by depletion of oil and gas reserves. Caio Koch-Weser, vice chair of Deutsche Bank, noted that "the Initiative shows in what dimensions and on what scale we must think if we are to master the challenges from climate change."

Even before this proposal, Algeria—for decades an oil exporter—was planning to build 6,000 megawatts of solar thermal generating capacity for export to Europe via undersea cable. The Algerians note that they have enough harnessable solar energy in their vast desert to power the entire world economy. This is not a mathematical error. A similar point often appears in the solar literature when it is noted that the sunlight striking the earth in one hour could power the world economy for one year. The German government was quick to respond to the Algerian initiative. The plan is to build a 1,900-mile high-voltage transmission line from Adrar deep in the Algerian desert to Aachen, a town on Germany's border with the Netherlands.

Although solar thermal power has been slow to get under way, utility-scale plants are being built rapidly now. The two leaders in this field are the United States and Spain. The United States has more than 40 solar thermal power plants operating, under construction, and under development that range from 10 to 1,200 megawatts each. Spain has 60 power plants in these same stages of development, most of which are 50 megawatts each.

One country ideally suited for CSP plants is India.

The Great Indian Desert in its northwest offers a huge opportunity for building solar thermal power plants. Hundreds of large plants in the desert could meet most of India's electricity needs. And because it is such a compact country, the distance for building transmission lines to major population centers is relatively short.

One of the attractions of utility-scale CSP plants is that heat during the day can be stored in molten salt at temperatures above 1,000 degrees Fahrenheit. The heat can then be used to keep the turbines running for eight or more hours after sunset.

The American Solar Energy Society notes that solar thermal resources in the U.S. Southwest can satisfy current U.S. electricity needs nearly four times over.

At the global level, Greenpeace, the European Solar Thermal Electricity Association, and the International Energy Agency's SolarPACES program have outlined a plan to develop 1.5 million megawatts of solar thermal power plant capacity by 2050. For Plan B we suggest a more immediate world goal of 200,000 megawatts by 2020, a goal that may well be exceeded as the economic potential becomes clearer.

The pace of solar energy development is accelerating as the installation of rooftop solar water heaters—the other use of solar collectors—takes off. China, for example, now has an estimated 1.9 billion square feet of rooftop solar thermal collectors installed, enough to supply 120 million Chinese households with hot water. With some 5,000 Chinese companies manufacturing these devices, this relatively simple low-cost technology has leapfrogged into villages that do not yet have electricity. For as little as $200, villagers can install a rooftop solar collector and take their first hot shower. This technology is sweeping China like wildfire, already approaching market saturation in some communities. Beijing's goal is to add another billion square feet to its rooftop solar water heating capacity by 2020, a goal it is likely to exceed.

Other developing countries such as India and Brazil may also soon see millions of households turning to this inexpensive water heating technology. Once the initial installment cost of rooftop solar water heaters is paid back, the hot water is essentially free.

In Europe, where energy costs are relatively high, rooftop solar water heaters are also spreading fast. In Austria, 15 percent of all households now rely on them for hot water. As in China, in some Austrian villages nearly all homes have rooftop collectors. Germany is also forging ahead. Some 2 million Germans are now living in homes where water and space are both heated by rooftop solar systems.

The U.S. rooftop solar water heating industry has historically concentrated on a niche market—selling and marketing 100 million square feet of solar water heaters for swimming pools between 1995 and 2005. Given this base, the industry was poised to mass-market residential solar water and space heating systems when federal tax credits were introduced in 2006. Led by Hawaii, California, and Florida, annual U.S. installation of these systems has more than tripled since 2005. The boldest initiative in the United States is California's goal of installing 200,000 solar water heaters by 2017. Not far behind is one launched in 2010 in New York State, which aims to have 170,000 residential solar water systems in operation by 2020.

Solar water and space heaters in Europe and China have a strong economic appeal, often paying for themselves from electricity savings in less than 10 years. With the cost of rooftop heating systems declining, many other countries will likely join Israel, Spain, and Portugal in mandating that all new buildings incorporate rooftop solar water heaters. The state of Hawaii requires that all new single-family homes have rooftop solar water heaters. Worldwide, Plan B calls for a total of 1,100 thermal gigawatts of rooftop solar water and space heating capacity by 2020.

The third principal component in the Plan B energy economy is geothermal energy. The heat in the upper six miles of the earth's crust contains 50,000 times as much energy as found in all of the world's oil and gas reserves combined—a startling statistic. Despite this abundance, as of mid-2010 only 10,700 megawatts of geothermal generating capacity have been harnessed worldwide, enough for some 10 million homes.

Roughly half the world's installed geothermal generating capacity is concentrated in the United States and the Philippines. Most of the remainder is generated in Mexico, Indonesia, Italy, and Japan. Altogether some 24 countries now convert geothermal energy into electricity. El Salvador, Iceland, and the Philippines respectively get 26, 25, and 18 percent of their electricity from geothermal power plants.

The geothermal potential to provide electricity, to heat homes, and to supply process heat for industry is vast. Among the geothermally rich countries are those bordering the Pacific in the so-called Ring of Fire, including Chile, Peru, Colombia, Mexico, the United States, Canada, Russia, China, Japan, the Philippines, Indonesia, and Australia. Other well-endowed countries include those along the Great Rift Valley of Africa, including Ethiopia, Kenya, Tanzania, and Uganda, and those around the Eastern Mediterranean. As of 2010, there are some 70 countries with projects under development or active consideration, up from 46 in 2007.

Beyond geothermal electrical generation, up to 100,000 thermal megawatts of geothermal energy are used directly—without conversion into electricity—to heat homes and greenhouses and to provide process heat to industry. For example, 90 percent of the homes in Iceland are heated with geothermal energy.

An interdisciplinary team of 13 scientists and engineers assembled by the Massachusetts Institute of Technology in 2006 assessed U.S. geothermal electrical generating potential. Drawing on the latest technologies, including those used by oil and gas companies in drilling and in enhanced oil recovery, the team estimated that enhanced geothermal systems could help the United States meet its energy needs 2,000 times over.

Even before this exciting new technology is widely deployed, investors are moving ahead with existing technologies. For many years, U.S. geothermal energy was confined largely to the Geysers project north of San Francisco, easily the world's largest geothermal generating complex, with 850 megawatts of generating capacity. Now the United States has more than 3,000 megawatts of existing geothermal electrical capacity and projects under development in 13 states. With California, Nevada, Oregon, Idaho, and Utah leading the way, and with many new companies in the field, the stage is set for a geothermal renaissance.

In mid-2008, Indonesia—a country with 128 active volcanoes and therefore rich in geothermal energy—announced that it would develop 6,900 megawatts of geothermal generating capacity; Pertamina, the state oil company, is responsible for developing the lion's share. Indonesia's oil production has been declining for the last decade, and in each of the last five years it has been an oil importer. As Pertamina shifts resources from oil to the development of geothermal energy, it could become the first oil company—state-owned or independent—to make the transition from oil to renewable energy.

Japan, which has 16 geothermal power plants with a total of 535 megawatts of generating capacity, was an early leader in this field. After nearly two decades of inactivity, this geothermally rich country—long known for its thousands of hot baths—is again building geothermal power plants.

Among the Great Rift countries in Africa, Kenya is the early geothermal leader. It now has 167 megawatts of generating capacity and is planning 1,200 more megawatts by 2015, enough to nearly double its current electrical generating capacity from all sources. It is aiming for 4,000 geothermal megawatts by 2030.

Beyond power plants, geothermal (ground source) heat pumps are now being widely used for both heating and cooling. These take advantage of the remarkable stability of the earth's temperature near the surface and then use that as a source of heat in the winter when the air temperature is low and a source of cooling in the summer when the air temperature is high. The great attraction of this technology is that it can provide both heating and cooling and do so with 25–50 percent less electricity than would be needed with conventional systems. In Germany, 178,000 ground-source heat pumps are now operating in residential or commercial buildings. At least 25,000 new pumps are installed each year.

Geothermal heat is ideal for greenhouses in northern countries. Russia, Hungary, Iceland, and the United States are among the many countries that use it to produce fresh vegetables in winter. With rising oil prices boosting fresh produce transport costs, this practice will likely become far more common.

If the four most populous countries located on the Pacific Ring of Fire—the United States, Japan, China, and Indonesia—were to seriously invest in developing their geothermal resources, it is easy to envisage a world with thousands of geothermal power plants generating some 200,000 megawatts of electricity, the Plan B goal, by 2020.

As oil and natural gas reserves are being depleted, the

world's attention is also turning to plant-based energy sources, including energy crops, forest industry byproducts, sugar industry byproducts, urban waste, livestock waste, plantations of fast-growing trees, crop residues, and urban tree and yard wastes—all of which can be used for electrical generation, heating, or the production of automotive fuels.

The potential use of energy crops is limited because even corn—the most efficient of the grain crops—can convert only 0.5 percent of solar energy into a usable form. In contrast, solar PV or solar thermal power plants convert roughly 15 percent of sunlight into electricity. And the value of electricity produced on an acre of land occupied by a wind turbine is over 300 times that of the corn-based ethanol produced on an acre. In this land–scarce world, energy crops cannot compete with solar-generated electricity, much less with wind power.

Yet another source of renewable energy is hydropower. The term has traditionally referred to dams that harnessed the energy in river flows, but today it also includes harnessing the energy in tides and waves as well as using smaller "in-stream" turbines to capture the energy in rivers and tides without building dams.

Roughly 16 percent of the world's electricity comes from hydropower, most of it from large dams. Some countries, such as Brazil, Norway, and the Democratic Republic of the Congo, get the bulk of their electricity from river power.

Tidal power holds a certain fascination because of its sheer potential scale. The first large tidal generating facility—La Rance Tidal Barrage, with a maximum generating capacity of 240 megawatts—was built 40 years ago in France and is still operating today. Within the last few years interest in tidal power has spread rapidly. South Korea, for example, is building a 254-megawatt project on its west coast that would provide all the electricity for the half-million people living in the nearby city of Ansan. At another site to the north, engineers are planning a 1,320-megawatt tidal facility in Incheon Bay, near Seoul. And New Zealand is planning a 200-megawatt project in the Kaipara Harbour on that country's northwest coast.

Wave power, though a few years behind tidal power, is also now attracting the attention of both engineers and investors. Scottish firms Aquamarine Power and SSE Renewables are teaming up to build 1,000 megawatts of wave and tidal power off the coast of Ireland and the United Kingdom. Ireland is planning 500 megawatts of wave generating capacity by 2020, enough to supply 8 percent of its electricity. Worldwide, the harnessing of wave power could generate a staggering 10,000 gigawatts of electricity, more than double current world electricity capacity from all sources.

We project that the 980 gigawatts (980,000 megawatts) of hydroelectric power in operation worldwide in 2009 will expand to 1,350 gigawatts by 2020. According to China's official projections, 180 gigawatts should be added there, mostly from large dams in the southwest. The remaining 190 gigawatts in our projected growth of hydropower would come from a scattering of large dams still being built in countries like Brazil and Turkey, dams now in the planning stages in sub-Saharan Africa, a large number of small hydro facilities, a fast-growing number of tidal projects, and numerous smaller wave power projects.

The efficiency gains outlined in the preceding chapter more than offset projected growth in energy use to 2020. The next step in the Plan B 80-percent reduction of carbon emissions comes from replacing fossil fuels with renewable sources of energy. In looking at the broad shifts from the reference year of 2008 to the Plan B energy economy of 2020, fossil-fuel-generated electricity drops by 90 percent worldwide as the fivefold growth in renewably generated electricity replaces all the coal and oil and 70 percent of the natural gas now used to generate electricity. Wind, solar photovoltaic, solar thermal, and geothermal will dominate the Plan B energy economy, but as noted earlier wind will be the centerpiece—the principal source of the electricity to heat, cool, and light buildings and to run cars and trains.

The Plan B projected tripling of renewable thermal heating generation by 2020, roughly half of it to come from direct uses of geothermal energy, will sharply reduce the use of both oil and gas to heat buildings and water. And in the transportation sector, energy use from fossil fuels drops by some 70 percent. This comes from shifting to all-electric and highly efficient plug-in hybrid cars that will run almost entirely on electricity, nearly all of it from renewable sources. And it also comes from shifting to electric trains, which are much more efficient than diesel-powered ones.

Each country's energy profile will be shaped by its unique endowment of renewable sources of energy. Some countries, such as the United States, Turkey, and China, will likely rely on a broad base of renewables—wind, solar, and geothermal power. But wind, including both onshore and offshore, is likely to emerge as the leading energy source in all three cases.

Other countries, including Spain, Algeria, Egypt, India, and Mexico, will turn primarily to solar thermal power plants and solar PV arrays to power their economies. For Iceland, Indonesia, Japan, and the Philippines, geothermal energy will likely be the mother lode. Still others will likely rely heavily on hydro, including Norway, Brazil, and Nepal. And some technologies, such as rooftop solar water heaters, will be used virtually everywhere.

As the transition progresses, the system for transporting energy from source to consumers will change beyond recognition. In the old energy economy, pipelines and tankers carried oil long distances from oil fields to consumers, including a huge fleet of tankers that moved oil from the Persian Gulf to markets on every continent. In the new energy economy, pipelines will be replaced by transmission lines.

The proposed segments of what could eventually become a national U.S. grid are beginning to fall into place. Texas is planning up to 2,900 miles of new transmission lines to link the wind-rich regions of west Texas and the Texas panhandle to consumption centers such as Dallas-Fort Worth and San Antonio. Two high-voltage direct current (HVDC) lines will link the rich wind resources of Wyoming and Montana to California's huge market. Other proposed lines will link wind in the northern Great Plains with the industrial Midwest.

In late 2009 Tres Amigas, a transmission company, announced its plans to build a "SuperStation" in Clovis, New Mexico, that would link the country's three major grids—the Western grid, the Eastern grid, and the Texas grid—for the first time. This would effectively create the country's first national grid. Scheduled to start construction in 2012 and to be completed in 2014, the SuperStation will allow electricity, much of it from renewable sources, to flow through the country's power transmission infrastructure.

Google made headlines when it announced in mid-October 2010 that it was investing heavily in a $5-billion offshore transmission project stretching from New York to Virginia, called the Atlantic Wind Connection. This will facilitate the development of enough offshore wind farms to meet the electricity needs of 5 million East Coast residents.

A strong, efficient national grid will reduce generating capacity needs, lower consumer costs, and cut carbon emissions. Since no two wind farms have identical wind profiles, each one added to the grid makes wind a more stable source of electricity. With the prospect of thousands of wind farms spread from coast to coast and a national grid, wind becomes a stable source of energy, part of baseload power.

Europe, too, is beginning to think seriously of investing in a supergrid. In early 2010, a total of 10 European companies formed Friends of the Supergrid, which is proposing to use HVDC undersea cables to build the European supergrid offshore, an approach that would avoid the time-consuming acquisition of land to build a continental land-based system. This grid could then mesh with the proposed Desertec initiative to integrate the offshore wind resources of northern Europe and the solar resources of North Africa into a single system that would supply both regions. The Swedish ABB Group, which in 2008 completed a 400-mile HVDC undersea cable linking Norway and the Netherlands, is well positioned to help build the necessary transmission lines.

Governments are considering a variety of policy instruments to help drive the transition from fossil fuels to renewables. These include tax restructuring, lowering the tax on income and raising the tax on carbon emissions to include the indirect costs of burning fossil fuels. If we can create an honest energy market, the transition to renewables will accelerate dramatically.

Another measure that will speed the energy transition is eliminating fossil fuel subsidies. At present, governments are spending some $500 billion per year subsidizing the use of fossil fuels. This compares with renewable energy subsidies of only $46 billion per year.

For restructuring the electricity sector, feed-in tariffs, in which utilities are required to pay set prices for electricity generated from renewable sources, have been remarkably successful. Germany's impressive early success with this measure has led to its adoption by some 50 other countries, including most of those in the European Union. In the United States, 29 states have adopted renewable portfolio standards requiring utilities to get up to 40 percent of their electricity from renewable sources. The United States has also used tax credits for wind, geothermal, solar photovoltaics, solar water and space heating, and ground-source heat pumps.

To achieve some goals, governments are simply using mandates, such as those requiring rooftop solar water heaters on all new buildings. Governments at all levels are adopting energy efficiency building codes. Each government has to select the policy instruments that work best in its particular economic and cultural setting.

In the new energy economy, our cities will be unlike any we have known during our lifetime. The air will be clean and the streets will be quiet, with only the scarcely audible hum of electric motors. Air pollution alerts will be a thing of the past as coal-fired power plants are dismantled and recycled and as gasoline- and-diesel-burning engines largely disappear.

This transition is now building its own momentum, driven by an intense excitement from the realization that we are tapping energy sources that can last as long as the earth itself. Oil wells go dry and coal seams run out, but for the first time since the Industrial Revolution, we are investing in energy sources that can last forever.

Data, endnotes, and additional resources can be found on Earth Policy's Web site, at www.earth-policy.org.

10

Restoring the Economy's Natural Support Systems

Pakistan's record flooding in the late summer of 2010 was the most devastating natural disaster in the country's history. The media coverage reported torrential rains as the cause, but there is much more to the story. When Pakistan was created in 1947, some 30 percent of the landscape was covered by forests. Now it is 4 percent. Pakistan's livestock herd outnumbers that of the United States. With little forest still standing and the countryside grazed bare, there was scant vegetation to retain the rainfall.

Pakistan, with 185 million people squeezed into an area only slightly larger than Texas, is an ecological basket case. If it cannot restore its forests and grazing lands, it will only suffer more "natural" disasters in the future. Pakistan's experience demonstrates all too vividly why restoration of the world's forests, grasslands, and soils is an integral part of Plan B. In this chapter we lay out both a plan for saving these natural support systems and a budget for doing so.

Restoring the earth will take an enormous international effort, one far more demanding than the Marshall Plan that helped rebuild war-torn Europe and Japan after World War II. And such an initiative must be undertaken at wartime speed before environmental deterioration translates into economic decline, just as it did for the

Sumerians, the Mayans, and many other early civilizations whose archeological sites we study today.

Protecting the 10 billion acres of remaining forests on earth and replanting many of those already lost, for example, are both essential for restoring the earth's health. Since 2000, the earth's forest cover has shrunk by 13 million acres each year, with annual losses of 32 million acres far exceeding the regrowth of 19 million acres.

Global deforestation is concentrated in the developing world. Tropical deforestation in Asia is driven primarily by the fast-growing demand for timber and increasingly by the expansion of oil palm plantations for fuel. In Latin America, in contrast, the fast-growing markets for soybeans and beef are together squeezing the Amazon. In Africa, the culprit is mostly fuelwood gathering and land clearing for agriculture.

Fortunately, there is a vast unrealized potential in all countries to lessen the various demands that are shrinking the earth's forest cover. In industrial nations, the greatest opportunity lies in reducing the amount of wood used to make paper. The use of paper, perhaps more than any other single product, reflects the throwaway mentality that evolved during the last century. The challenge is to replace facial tissues, paper napkins, and paper shopping bags with reusable cloth alternatives.

The goal is first to reduce paper use and then to recycle as much as possible. The rates of paper recycling in the top 10 paper-producing countries range widely—from China and Finland on the low end, recycling less than 40 percent of the paper they use, to Japan and Germany on the higher end, each between 70 and 80 percent, and South Korea, which recycles an impressive 91 percent. The United States, the world's largest paper consumer, is far behind the leaders, but it has raised the share of paper recycled from roughly 20 percent in 1980 to 59 percent in 2009. If every country recycled as much of its paper as South Korea does, the amount of wood pulp used to produce paper worldwide would drop by more than one third.

In developing countries, the focus needs to be on reducing fuelwood use. Indeed, fuelwood accounts for just over half of all wood removed from the world's forests. Some international aid agencies, including the U.S. Agency for International Development, are sponsoring fuelwood efficiency projects. In September 2010, the United Nations Foundation, leading a coalition of groups, announced plans to get 100 million more-efficient stoves into homes by 2020. Highly efficient cookstoves not only use far less wood than traditional stoves, they also pollute less. Over the longer term, pressure on forests can be reduced by replacing firewood with solar thermal cookers or even with electric hotplates powered with renewable energy.

Another challenge is to harvest forests responsibly. There are two basic approaches to timber harvesting. One is clearcutting. This practice is environmentally devastating, leaving eroded soil and silted streams, rivers, and irrigation reservoirs in its wake. The alternative is simply to cut only mature trees on a selective basis, leaving the forest largely intact. This ensures that forest productivity can be maintained in perpetuity.

Forest plantations can reduce pressures on the earth's remaining forests as long as they do not replace old-growth forest. As of 2010, the world had 652 million acres in planted forests, more than one third as much land as is planted in grain. Tree plantations produce mostly wood for paper mills or for wood reconstitution mills. Increasingly, reconstituted wood is substituted for natural wood as lumber and construction industries adapt to a shrinking supply of large logs from natural forests.

Six countries account for 60 percent of productive tree plantations. China, which has little original forest remaining, is by far the largest, with 134 million acres. India and the United States follow, with 42 million acres each. Russia, Canada, and Sweden are close behind. As tree farming expands, it is starting to shift geographically to the moist tropics, where yields are much higher. In eastern Canada, one hectare (2.47 acres) of forest plantation produces 4 cubic meters of wood per year. In the southeastern United States, the yield is 10 cubic meters. But in Brazil, newer plantations are getting close to 40 cubic meters.

The U.N. Food and Agriculture Organization projects that as plantation area expands and yields rise, the harvest could more than triple between 2005 and 2030. It is entirely conceivable that plantations could one day satisfy most of the world's demand for industrial wood, thus helping protect the world's remaining natural forests.

Planting trees on degraded or disturbed land not only reduces soil erosion, it also helps pull carbon dioxide (CO_2) out of the atmosphere. In recent years, the shrinkage of forests in tropical regions has released 2.2 billion tons of carbon into the atmosphere annually. Meanwhile, expanding forests in the temperate regions are absorbing close to 700 million tons of carbon. On balance, therefore, some 1.5 billion tons of carbon are released into the atmosphere each year from forest loss, roughly one fourth as much as from fossil fuel burning.

The Plan B goals are to end net deforestation worldwide and to sequester carbon through a variety of tree planting initiatives and the adoption of improved agricultural land management practices. Although banning deforestation may seem far-fetched, environmental damage has pushed Thailand, the Philippines, and China to implement partial or complete bans on logging. All three bans followed devastating floods and mudslides resulting from the loss of forest cover.

In China, after suffering record losses from weeks of

nonstop flooding in the Yangtze River basin in 1998, the government noted that when forest policy was viewed not through the eyes of the individual logger but through those of society as a whole, it simply did not make economic sense to continue deforesting. The flood control service of trees standing, they said, was three times as valuable as the timber from trees cut.

Protecting the earth's soil also warrants a worldwide ban on the clearcutting of forests in favor of selective harvesting, simply because each successive clearcut brings heavy soil loss and eventual forest degeneration. Restoring the earth's tree and grass cover, as well as practicing conservation agriculture, protects soil from erosion, reduces flooding, and sequesters carbon.

International environmental groups such as Greenpeace and WWF have negotiated agreements to halt deforestation in the Brazilian Amazon and in parts of Canada's boreal forests. Daniel Nepstad and colleagues reported in *Science* in 2009 on two recent developments that together may halt deforestation in the Amazon basin. One is Brazil's Amazon deforestation reduction target that was announced in 2008, which prompted Norway to commit $1 billion if there is progress toward this goal. The second is a marketplace transition in the beef and soy industries to avoid Amazon deforesters in their supply chains.

If Brazil's Amazon rainforest continues to shrink, it may also continue to dry out, becoming vulnerable to fire. If this rainforest were to disappear, it would likely be replaced largely by desert and scrub forestland. The reduced capacity of the rainforest to cycle water to the interior of the continent would threaten the agricultural areas in the west and south.

Recognizing the central role of forests in modulating climate, the Intergovernmental Panel on Climate Change has examined the potential for tree planting and improved forest management to sequester CO_2. Since every newly planted tree seedling in the tropics removes an average of 50 kilograms of CO_2 from the atmosphere each year during its growth period of 20–50 years, compared with 13 kilograms of CO_2 per year for a tree in the temperate regions, much of the afforestation and reforestation opportunity is found in tropical countries.

What is needed is a tree planting effort to both conserve soil and sequester carbon. To achieve these goals, billions of trees need to be planted on millions of acres of degraded lands that have lost their tree cover and on marginal croplands and pasturelands that are no longer productive.

This global forestation plan to remove atmospheric CO_2 would need to be funded by the industrial countries that put most of it there. In comparison with other mitigation strategies, stopping deforestation and planting trees are relatively inexpensive. They pay for themselves many times over. An independent body could be set up to administer and monitor the vast tree planting initiative. The key is moving quickly to stabilize climate before temperature rises too high, thus giving these trees the best possible chance of survival.

There are already many tree planting initiatives proposed or under way. Kenya's Nobel laureate, Wangari Maathai, who years ago organized women in Kenya and several nearby countries to plant 30 million trees, inspired the Billion Tree Campaign that is managed by the U.N. Environment Programme. The initial goal in 2006 was to plant 1 billion trees. If half of those trees survive, they will sequester 5.6 million tons of carbon per year. By the end of 2009, over 10 billion trees had been planted.

Some state and provincial governments have also joined in. Uttar Pradesh, India's most populous state, mobilized 600,000 people to plant 10.5 million trees in a single day in July 2007, putting the trees on farmland, in state forests, and on school grounds. Since then, India has planted 2 billion additional trees. China, which has planted 2.9 billion trees, is now the leader in the Billion Tree Campaign. Among the other leaders in this initiative are Ethiopia, with 1.5 billion trees, and Turkey, with over 700 million trees planted.

Some countries reforest on their own. South Korea is in many ways a reforestation model for the rest of the world in this respect. When the Korean War ended half a century ago, the mountainous country was largely deforested, much as Haiti is today. Beginning around 1960, under the dedicated leadership of President Park Chung Hee, the South Korean government launched a national reforestation effort. Hundreds of thousands of people were mobilized in village cooperatives to dig trenches and to create terraces for supporting trees on barren mountains. Se-Kyung Chong, a researcher at the Korea Forest Research Institute, notes that "the result was a seemingly miraculous rebirth of forests from barren land."

Today forests cover nearly 65 percent of the country, an area of more than 15 million acres. While driving across South Korea in November 2000, it was gratifying to see the luxuriant stands of trees on mountains that a generation earlier were bare. We can reforest the earth!

Planting trees is just one of many activities that will remove meaningful quantities of carbon from the atmosphere. Improved grazing and land management practices that increase the organic matter content in soil also sequester carbon.

The 1930s Dust Bowl that threatened to turn the U.S. Great Plains into a vast desert was a traumatic experience that led to revolutionary changes in American agricultural practices, including the planting of tree shelterbelts (rows of trees planted beside fields to slow wind and thus reduce wind erosion) and strip cropping (the planting of wheat on alternate strips with fallowed land each year). Strip cropping permits soil moisture to accumulate on the fallowed strips, while the alternating planted strips reduce wind speed and hence erosion on the idled land.

In 1985, the U.S. Department of Agriculture, with strong support from the environmental community, created the Conservation Reserve Program (CRP) to reduce soil erosion and control overproduction of basic commodities. By 1990 there were some 35 million acres of highly erodible land with permanent vegetative cover under 10-year contracts. Under this program, farmers were paid to plant fragile cropland in grass or trees. The retirement of those 35 million acres under the CRP, together with the use of conservation practices on 37 percent of all cropland, reduced annual U.S. soil erosion from 3.1 billion tons to 1.9 billion tons between 1982 and 1997. The U.S. approach offers a model for the rest of the world.

Another tool in the soil conservation toolkit is conservation tillage, which includes both no-till and minimum tillage. Instead of the traditional cultural practices of plowing land and discing or harrowing it to prepare the seedbed, and then using a mechanical cultivator to control weeds in row crops, farmers simply drill seeds directly through crop residues into undisturbed soil, controlling weeds with herbicides. The only soil disturbance is the narrow slit in the soil surface where the seeds are inserted, leaving the remainder of the soil covered with crop residue and thus resistant to both water and wind erosion. In addition to reducing erosion, this practice retains water, raises soil carbon content, and greatly reduces energy use for tillage.

In the United States, the no-till area went from 17 million acres in 1990 to 65 million acres in 2007. Now widely used in the production of corn and soybeans, no-till has spread rapidly, covering 63 million acres in Brazil and

Argentina and 42 million in Australia. Canada, not far behind, rounds out the five leading no-till countries. Farming practices that reduce soil erosion and raise cropland productivity such as minimum-till, no-till, and mixed crop-livestock farming usually also lead to higher soil carbon content and soil moisture. In Kazakhstan, the 3 million acres in no-till seemed to fare better than land in conventional farming during the great Russian heat wave and drought of 2010.

In sub-Saharan Africa, where the Sahara is moving southward all across the Sahel, countries are concerned about the growing displacement of people as grasslands and croplands turn to desert. As a result, the African Union has launched the Green Wall Sahara Initiative. This plan, originally proposed in 2005 by Olusegun Obasanjo when he was president of Nigeria, calls for planting a 4,300-mile band of trees, 9 miles wide, stretching across Africa from Senegal to Djibouti. Senegal, which is losing 124,000 acres of productive land each year and which would anchor the green wall on the western end, has planted 326 miles of the band. A $119-million grant from the Global Environment Facility in June 2010 gave the project a big boost. Senegal's Environment Minister, Modou Fada Diagne, says, "Instead of waiting for the desert to come to us, we need to attack it." One key to the success of this initiative is improving management practices, such as rotational grazing.

In the end, the only viable way to eliminate overgrazing on the two fifths of the earth's land surface classified as rangelands is to reduce the size of flocks and herds. Not only do the excessive numbers of cattle, sheep, and goats remove the vegetation, but their hoofs pulverize the protective crust of soil that is formed by rainfall and that naturally checks wind erosion. In some situations, the preferred option is to keep the animals in restricted areas, bringing the forage to them. India, which has successfully adopted this practice to build the world's largest dairy industry, is a model for other countries.

Oceanic fisheries, another major source of animal protein, are also under intense pressure. For decades, governments have tried to save specific fisheries by restricting the catch of individual species. Sometimes this worked; sometimes it failed and fisheries collapsed. In recent years, support for another approach—the creation of marine reserves or marine parks—has been gaining momentum. These reserves, where fishing is banned, serve as natural hatcheries, helping to repopulate the surrounding area.

In 2002, at the World Summit on Sustainable Development in Johannesburg, coastal nations pledged to create national networks of marine reserves or parks that would cover 10 percent of the world's oceans by 2012. Together these could constitute a global network of such parks.

Progress is slow. Today some 5,000 marine protected areas cover less than 1 percent of the world's oceans. Even more distressing, fishing is banned in only 12.8 percent of those areas. And a survey of 255 marine reserves reported that only 12 were routinely patrolled to enforce the ban.

In 2001 Jane Lubchenco, former President of the American Association for the Advancement of Science and now head of the National Oceanic and Atmospheric Administration, released a statement signed by 161 leading marine scientists calling for urgent action to create the global network of marine reserves. Drawing on the research on scores of marine parks, she said: "All around the world there are different experiences, but the basic message is the same: marine reserves work, and they work fast. It is no longer a question of whether to set aside fully protected areas in the ocean, but where to establish them."

Sea life improves quickly once the reserves are established. A case study of a snapper fishery off the coast of New England showed that fishers, though they violently opposed the establishment of the reserve, now champion it because they have seen the local population of snapper increase 40-fold. In the Gulf of Maine, all fishing methods that put groundfish at risk were banned within three marine reserves totaling 6,600 square miles. Unexpectedly, scallops flourished in this undisturbed environment, and their populations increased by up to 14-fold within five years. This buildup within the reserves also greatly increased the scallop population outside the reserves. Within a year or two of establishing a marine reserve, population densities increased 91 percent, average fish size went up 31 percent, and species diversity rose 20 percent.

But the challenges we face are changing, and so must the response. The traditional approach to protecting biological diversity by building a fence around an area and calling it a park or nature preserve is no longer sufficient. If we cannot also stabilize population and climate, there is not an ecosystem on earth that we can save, no matter how high the fence.

We can roughly estimate how much it will cost to reforest the earth, protect topsoil, restore rangelands and fisheries, stabilize water tables, and protect biological diversity. The goal is not to offer a set of precise numbers but rather to provide a set of reasonable estimates for an earth restoration budget.

In calculating reforestation costs, the focus is on developing countries, since forested area is already expanding in the northern hemisphere's industrial countries. Meeting the growing fuelwood demand in developing countries will require an estimated 140 million additional acres of forested area. Conserving soils and restoring hydrological stability would require another 250 million acres in thousands of watersheds in developing countries. Recognizing some overlap between these two, we will reduce the total to 380 million acres. Beyond this, an additional 75 million acres will be needed to produce lumber, paper, and other forest products.

Only a small share of the tree planting will likely come from plantations. Much of it will be on the outskirts of villages, along field boundaries and roads, on small plots of marginal land, and on denuded hillsides. The labor for this will be local; some will be paid labor, some volunteer. Much of it will be rural off-season labor.

If seedlings cost $40 per thousand, as the World Bank estimates, and if the typical planting density is roughly 800 per acre, then seedlings cost $32 per acre. Labor costs for planting trees are high, but since much of the labor would consist of locally mobilized volunteers, we are assuming a total of $160 per acre, including both seedlings and labor. With a total of 380 million acres to be planted over the next decade, this will come to roughly 38 million acres per year at $160 each for an annual expenditure of $6 billion.

Planting trees to conserve soil, reduce flooding, and provide firewood sequesters carbon. But because climate stabilization is essential, we tally the cost of planting trees for carbon sequestration separately. Doing so would reforest or afforest hundreds of millions of acres of marginal lands over 10 years. Because it would be a more commercialized undertaking focused exclusively on wasteland reclamation and carbon sequestration, it would be more costly. Assuming a value of sequestered carbon of $200 per ton, it would cost close to $17 billion per year.

Conserving the earth's topsoil by reducing erosion to the rate of new soil formation or below has two parts. One is to retire the highly erodible land that cannot sustain cultivation—the estimated one tenth of the world's cropland that accounts for perhaps half of all excess erosion. For the United States, that has meant retiring nearly 35 million acres. The cost of keeping this land out of production is close to $50 per acre. In total, annual payments to farmers to plant this land in grass or trees under 10-year contracts approaches $2 billion.

In expanding these estimates to cover the world, it is assumed that roughly 10 percent of the world's cropland is highly erodible, as in the United States, and should be planted in grass or trees before the topsoil is lost and it becomes barren land. In both the United States and China, which together account for 40 percent of the world grain harvest, the official goal is to retire one tenth of all cropland. For the world as a whole, converting 10 percent of cropland that is highly erodible to grass or trees seems like a reasonable goal. Since this costs roughly $2 billion in the United States, which has one eighth of the world's cropland, the total for the world would be $16 billion annually.

The second initiative on topsoil consists of adopting conservation practices on the remaining land that is subject to excessive erosion—that is, erosion that exceeds the natural rate of new soil formation. This initiative includes incentives to encourage farmers to adopt conservation practices such as contour farming, strip cropping, and, increasingly, minimum-till or no-till farming. These expenditures in the United States total roughly $1 billion per year.

Assuming that the need for erosion control practices elsewhere is similar to that in the United States, we again multiply the U.S. expenditure by eight to get a total of $8 billion for the world as a whole. The two components together—$16 billion for retiring highly erodible land and $8 billion for adopting conservation practices—give an annual total for the world of $24 billion.

For cost data on rangeland protection and restoration, we turn to the U.N. Plan of Action to Combat Desertification. This plan, which focuses on the world's dryland regions, containing nearly 90 percent of all rangeland, estimates that it would cost roughly $183 billion over a 20-year restoration period—or $9 billion per year. The key restoration measures include improved rangeland management, financial incentives to eliminate overstocking, and revegetation with appropriate rest periods, during which grazing would be banned.

This is a costly undertaking, but every $1 invested in rangeland restoration yields a return of $2.50 in income from the increased productivity of the earth's rangeland ecosystem. From a societal point of view, countries with large pastoral populations where the rangeland deterioration is concentrated are invariably among the world's poorest. The alternative to action—ignoring the deterioration—brings a loss not only of land productivity but also of livelihood, and ultimately leads to millions of refugees. Restoring vulnerable land will also have carbon sequestration benefits.

For restoring fisheries, a U.K. team of scientists led by Andrew Balmford of the Conservation Science Group at Cambridge University has analyzed the costs of operating marine reserves on a large scale based on data from 83 relatively small, well-managed reserves. They concluded that managing reserves that covered 30 percent of the world's oceans would cost $12–14 billion a year. But this did not take into account the likely additional income from recovering fisheries, which would reduce the actual cost.

At stake in the creation of a global network of marine reserves is not just the protection of fisheries but also a possible increase in the annual oceanic fish catch worth $70–80 billion. Balmford said, "Our study suggests that we could afford to conserve the seas and their resources in perpetuity, and for less than we are now spending on subsidies to exploit them unsustainably." The creation of the global network of marine reserves—"Serengetis of the seas," as some have dubbed them—would also create more than 1 million jobs.

In many countries, the capital needed to fund a program to raise water productivity can come from eliminating subsidies that often encourage the wasteful use of irrigation water. Sometimes these are energy subsidies for irrigation, as in India; other times they are subsidies that provide water at prices well below costs, as in the United States. Removing these subsidies will effectively raise the price of water, thus encouraging its more efficient use. In terms of additional resources needed worldwide, including research needs and the economic incentives for farmers, cities, and industries to use more water-efficient practices and technologies, we assume it will take an additional annual expenditure of $10 billion.

For wildlife protection, the World Parks Congress estimates that the annual shortfall in funding needed to manage and protect existing areas designated as parks comes to roughly $25 billion a year. Additional areas needed, including those encompassing the biologically diverse hotspots not yet included in designated parks, would cost perhaps another $6 billion a year, yielding a total of $31 billion.

Altogether, then, restoring the economy's natural support systems—reforesting the earth, protecting topsoil, restoring rangelands and fisheries, stabilizing water tables, and protecting biological diversity—will require additional expenditures of just $110 billion per year. Many will ask, Can the world afford these investments? But the only appropriate question is, Can the world afford the consequences of not making these investments?

Data, endnotes, and additional resources can be found on Earth Policy's Web site, at www.earth-policy.org.

11

Eradicating Poverty, Stabilizing Population, and Rescuing Failing States

In 1974, Miguel Sabido, a vice president of Televisa, Mexico's national television network, ran a series of soap opera segments on illiteracy. The day after one of the characters visited a literacy office wanting to learn how to read and write, a quarter-million people showed up at these offices in Mexico City. Eventually 840,000 Mexicans enrolled in literacy courses after watching the series.

While many analysts focus on the role of formal education in social change, soap operas on radio and television can quickly change people's attitudes about literacy, reproductive health, and family size. A well-written soap opera can have a profound near-term effect on population growth. It costs relatively little and can proceed while formal educational systems are being expanded.

Sabido, a pioneer in this exciting new option for raising awareness, dealt with contraception in a soap opera entitled *Acompáñame*, which translates as *Come With Me*. Over the span of a decade this drama series helped reduce Mexico's birth rate by 34 percent.

Other groups outside Mexico quickly picked up his approach. The U.S.-based Population Media Center (PMC), headed by William Ryerson, has initiated projects in some 17 countries and plans to launch projects in several others. The PMC's work in Ethiopia over the last several years provides a telling example. Its radio serial dramas broadcast in Amharic and Oromiffa have addressed issues of reproductive health and gender equity.

A survey two years after the broadcasts began in 2002 found that 63 percent of new clients seeking reproductive health care at Ethiopia's 48 service centers had listened to a PMC drama. There was a 55-percent increase in family planning use among married women in the Amhara region of Ethiopia who listened to these dramas. The average number of children per woman in the region dropped from 5.4 to 4.3. This is an exciting result because reduction in family size makes it easier to eradicate poverty. And conversely, eradicating poverty accelerates the shift to smaller families.

Poverty has many faces, such as hunger, illiteracy, and low life expectancy. In 2005, nearly 1.4 billion people around the world were living on less than $1.25 a day, which the World Bank classifies as extreme poverty. The highest regional concentration of poverty is in sub-Saharan Africa, where extreme poverty afflicts over half the sub-continent's 863 million people. Among the world's failing or fragile states, poverty is pervasive, also affecting more than half the population. Yet unlike in sub-Saharan Africa, where some progress (albeit slow) has been made, the prospects for alleviating poverty in failing states look pretty grim without major state rehabilitation.

Since those living in poverty spend a large share of their income on food, it came as no surprise when the World Bank reported in early 2009 that between 2005 and 2008 the ranks of the poor expanded by at least 130 million people because of higher food prices. The Bank also observed that 44 million more children may suffer permanent cognitive and physical injury caused by the rise in malnutrition. The effect of rising food prices was then compounded by the global economic crisis, which dramatically expanded the number of unemployed and reduced the flow of remittances from family members working abroad.

Although as of 2010 the world economy is starting to recover, the recession's setbacks to eradicating poverty are likely to persist for some years. Hunger and disease are on the march in many parts of the world, partly offsetting gains made in countries like China and Brazil. The late twentieth century's decline in hunger and malnourishment was reversed in 1996—rising from 788 million to 833 million in 2001, passing 900 million in 2008, to over 1 billion in 2009.

Eradicating poverty is the key to stabilizing population, improving food security, and minimizing state failure. There are many success stories of people moving up the economic ladder, but none are as impressive as China's. There, a fast-growing economy and a continued shift to small families dropped the number of Chinese living in extreme poverty from 683 million in 1990 to 208 million in 2005. The share of the population living in extreme poverty plummeted from 60 percent to 16 percent.

Brazil has also succeeded in sharply reducing poverty through its Bolsa Familia program, an effort initiated by President Luiz Inácio Lula da Silva in 2003. This program offers poor mothers up to $35 a month as long as they keep their children in school, have them vaccinated, and make sure they get regular physical checkups. Between 1990 and 2005, the share of the population living in extreme poverty dropped from 15 to 5 percent. Serving over 12 million households, nearly one fourth of the country's population, the program raised incomes among the poor by 22 percent over a five-year span, while incomes among the rich rose by only 5 percent. The gap between rich and poor is itself a source of instability. Brazil's success in reducing that gap is remarkable because, as Rosani Cunha, the program's former director in Brasilia, observed, "There are very few countries that reduce inequality and poverty at the same time."

Children without any formal education start life with a severe handicap, one that almost ensures they will remain in abject poverty and that the gap between the poor and the rich will continue to widen. So another key to eradicating poverty is to make sure that all children have at least a primary school education. Nobel Prize–winning economist Amartya Sen asserts that "illiteracy and innumeracy are a greater threat to humanity than terrorism."

The world is at least making progress on the education front. The number of elementary-school-aged children who were not in school dropped encouragingly from 106 million in 1999 to 69 million in 2008. And by 2005, almost two thirds of developing countries had reached another basic educational goal: gender parity in elementary school enrollment. This is not only a landmark achievement in its own right, it is also a key to stabilizing population. As female educational levels rise, fertility falls. Economist Gene Sperling notes that a study of 72 countries found that "the expansion of female secondary education may be the single best lever for achieving substantial reductions in fertility."

The goal of reducing illiteracy must extend beyond the elementary level. As the world becomes ever more integrated economically, its nearly 800 million illiterate adults are severely handicapped. We can overcome this deficit by launching adult literacy programs that rely heavily on volunteers. The international community can contribute by offering seed money to provide educational materials and outside advisors where needed. Bangladesh and Iran, both of which have successful adult literacy programs, can serve as models. An adult literacy program would add $4 billion per year to the budget to save civilization.

The World Bank has taken the lead in seeking universal primary education with its Education for All fast-track initiative, where any country with a well-designed plan to achieve universal primary education is eligible for Bank financial support. The three principal requirements are that the country submit a sensible plan to reach universal basic education, commit a meaningful share of its own resources to the plan, and have transparent budgeting and accounting practices. If fully implemented, all children in poor countries would get a primary school education by 2015, helping them to break out of poverty. An estimated $10 billion in external financing, beyond what is being spent today, is needed to achieve this.

Few incentives to get children into school are as effective as a school lunch program, especially in the poorest countries. Children who are ill or hungry miss many days of school. And even when they can attend, they do not learn as well. Economist Jeffrey Sachs notes, "Sick children often face a lifetime of diminished productivity because of interruptions in schooling together with cognitive and physical impairment." But when school lunch programs are launched in low-income countries, school enrollment jumps, academic performance goes up, and children spend more years in school.

Girls, who are more often expected to work at home, especially benefit. Particularly where programs include take-home rations, school meals lead to girls staying in school longer, marrying later, and having fewer children. This is a win-win-win situation. Launching school lunch programs to reach the 66 million youngsters who currently go to school hungry would cost an estimated $3 billion per year beyond what the U.N. World Food Programme is now spending to reduce hunger.

If children are to benefit from school lunch programs, we must improve nutrition before children even get to school age. Former Senator George McGovern suggests that a WIC program (for women, infants, and children) that provides nutritious food to needy pregnant and nursing mothers, similar to a program he helped launch in the United States, should be available in poor countries. Based on 33 years of U.S. experience, it is clearly successful at improving the nutrition, health, and development of preschool children from low-income families. If the program were expanded to reach pregnant women, nursing mothers, and small children in the 44 poorest countries, it would help eradicate hunger among millions of small children. And it would require additional expenditures of only $4 billion per year.

Ensuring access to a safe and reliable supply of water for the estimated 884 million people who lack it is essential to better health for all and a key to reducing infant mortality. Since clean water reduces the incidence of diarrheal and parasitic diseases, it also curbs nutrient loss and malnutrition. A realistic option in many developing-world cities is to bypass efforts to build costly water-based sewage removal and treatment systems and to opt instead for water-free waste disposal systems, including the increasingly popular odorless dry-compost toilets that do not disperse disease pathogens. This switch would simultaneously help alleviate water scarcity, reduce the spread of disease agents in water systems, and help close the nutrient cycle—another win-win-win situation.

Additional investments can help the many countries that cannot afford vaccines for childhood diseases and are falling behind in their vaccination programs. Lacking the funds to invest today, these countries will pay a far higher price tomorrow. In an effort to fill this funding gap, the Bill and Melinda Gates Foundation announced in early 2010 that it would provide over $10 billion this decade "to help research, develop, and deliver vaccines to the world's poorest countries."

More broadly, a World Health Organization study analyzing the economics of health care in developing countries concluded that providing the most basic health care services, the sort that could be supplied by a village-level clinic, would yield enormous economic benefits. The authors estimate that providing basic universal health care in developing countries will require donor grants totaling on average $33 billion a year through 2015. This figure includes funding for the Global Fund to Fight AIDS, Tuberculosis and Malaria and for universal childhood vaccinations.

When it comes to population growth, the United Nations has three primary projections. The medium projection, the one most commonly used, has world population reaching 9.2 billion by 2050. The high one reaches 10.5 billion. The low projection, which assumes that the world will quickly move below replacement-level fertility, has population peaking at 8 billion in 2042 and then declining. If the goal is to eradicate poverty, hunger, and illiteracy, then we have little choice but to strive for the lower projection.

Slowing world population growth means ensuring that all women who want to plan their families have access to family planning services. Unfortunately, this is currently not the case for 215 million women, 59 percent of whom live in sub-Saharan Africa and the Indian subcontinent. These women and their families represent roughly 1 billion of the earth's poorest residents, for whom unintended pregnancies and unwanted births are an enormous burden. Former U.S. Agency for International Development (AID) official J. Joseph Speidel notes that "if you ask anthropologists who live and work with poor people at the village level... they often say that women live in fear of their next pregnancy. They just do not want to get pregnant."

The United Nations Population Fund and the

Guttmacher Institute estimate that meeting the needs of these 215 million women who lack reproductive health care and effective contraception could each year prevent 53 million unwanted pregnancies, 24 million induced abortions, and 1.6 million infant deaths. Along with the provision of additional condoms needed to prevent HIV and other sexually transmitted infections, a universal family planning and reproductive health program would cost an additional $21 billion in funding from industrial and developing countries.

The good news is that governments can help couples reduce family size very quickly when they commit to doing so. My colleague Janet Larsen writes that in just one decade Iran dropped its near-record population growth rate to one of the lowest in the developing world.

When Ayatollah Khomeini assumed leadership in Iran in 1979 and launched the Islamic revolution, he immediately dismantled the well-established family planning programs and instead advocated large families. At war with Iraq between 1980 and 1988, Khomeini wanted to increase the ranks of soldiers for Islam. His goal was an army of 20 million.

Fertility levels climbed in response to his pleas, pushing Iran's annual population growth to a peak of 4.2 percent in the early 1980s, a level approaching the biological maximum. As this enormous growth began to burden the economy and the environment, the country's leaders realized that overcrowding, environmental degradation, and unemployment were undermining Iran's future.

In 1989 the government did an about-face and restored its family planning program. In May 1993, a national family planning law was passed. The resources of several government ministries, including education, culture, and health, were mobilized to encourage smaller families. Iran Broadcasting was given responsibility for raising awareness of population issues and of the availability of family planning services. Some 15,000 "health houses" or clinics were established to provide rural populations with health and family planning services.

Religious leaders were directly involved in what amounted to a crusade for smaller families. Iran introduced a full panoply of contraceptive measures, including the option of vasectomy—a first among Muslim countries. All forms of birth control, including the pill and sterilization, were free of charge. Iran even became the only country to require couples to take a course on modern contraception before receiving a marriage license.

In addition to the direct health care interventions, Iran also launched a broad-based effort to raise female literacy, boosting it from 25 percent in 1970 to more than 70 percent in 2000. Female school enrollment increased from 60 to 90 percent. Television was used to disseminate information on family planning throughout the country, taking advantage of the 70 percent of rural households with TV sets. As a result of this initiative, family size in Iran dropped from seven children to fewer than three. From 1987 to 1994, Iran cut its population growth rate by half, an impressive achievement that shows how a full-scale mobilization of society can accelerate the shift to smaller families.

The bad news is that in July 2010 Iranian President Mahmoud Ahmadinejad declared the country's family planning program ungodly and announced a new pronatalist policy. The government would pay couples to have children, depositing money in each child's bank account until age 18. The effect of this new program on Iran's population growth remains to be seen.

Shifting to smaller families brings generous economic dividends. In Bangladesh, for example, analysts concluded that $62 spent by the government to prevent an unwanted birth saved $615 in expenditures on other social services. For donor countries, ensuring that men and women everywhere have access to the services they need would yield strong social returns in improved education and health care. Put simply, the costs to society of not filling the family planning gap may be greater than we can afford.

Many developing countries in Asia, Africa, and Latin America were successful in quickly reducing their fertility within a generation or so after public health and medical gains lowered their mortality rates. Among these were Brazil, Chile, China, South Korea, Thailand, and Turkey. But many others did not follow this path and have been caught in the demographic trap—including Afghanistan, Ethiopia, Iraq, Nigeria, Pakistan, and Yemen.

Slowing population growth brings with it what economists call the demographic bonus. When countries move quickly to smaller families, growth in the number of young dependents—those who need nurturing and educating—declines relative to the number of working adults. At the individual level, removing the financial burden of large families allows more people to escape from poverty. At the national level, the demographic bonus causes savings and investment to climb, productivity to surge, and economic growth to accelerate.

Japan, which cut its population growth in half between 1951 and 1958, was one of the first countries to benefit from the demographic bonus. South Korea and Taiwan followed, and more recently China, Thailand, and Viet Nam have been helped by earlier sharp reductions in birth rates. Although this effect lasts for only a few decades, it is usually enough to launch a country into the modern era. Indeed, except for a few oil-rich countries, no developing country has successfully modernized without slowing population growth.

Countries that do not succeed in reducing fertility early on face the compounding of 3 percent growth per year or 20-fold per century. Such rapid population growth can easily strain limited land and water resources. With large "youth bulges" outrunning job creation, the growing number of unemployed young men increases the risk of conflict. This also raises the odds of becoming a failing state. One of the leading challenges facing the international community is how to prevent that slide into chaos. Continuing with business as usual with international assistance programs is not working. The stakes could not be higher. Somehow we must turn the tide of state decline.

Some donor countries have recognized that failing states need special attention. Since state failure is, by its nature, systemic, a systemic response is called for—one that is responsive to the many interrelated sources of failure. Traditional, project-oriented development assistance is not likely to reverse state failure. Rather, it requires a much deeper, across-the-board engagement with the failing state.

Reversing the process of state failure is a much more challenging, demanding process than anything the international community has faced, including the rebuilding of war-torn states after World War II. And it requires a level of interagency cooperation that no donor country has yet achieved. Pauline H. Baker, President of the Fund for Peace, suggests that a major stumbling block is that industrial governments do not recognize state failure as an entirely new kind of problem and thus do not design a comprehensive, integrated strategy to combat it.

Within the U.S. government, efforts to deal with weak and failing states are fragmented. Several departments are involved, including State, Treasury, and Agriculture. And within the State Department, several different offices are concerned with this issue. This lack of focus was recognized by the Hart-Rudman U.S. Commission on National Security in the Twenty-first Century:

"Responsibility today for crisis prevention and response is dispersed in multiple AID and State bureaus, and among State's Under Secretaries and the AID Administrator. In practice, therefore, no one is in charge."

What is needed now is a new cabinet-level agency—a Department of Global Security (DGS)—that would fashion a coherent policy toward each weak and failing state. This recommendation, initially set forth in a report of the Commission on Weak States and U.S. National Security, recognizes that the threats to security are now coming less from military power and more from the trends that undermine states, such as rapid population growth, poverty, deteriorating environmental support systems, and spreading water shortages. The new agency would incorporate AID (now part of the State Department) and all the various foreign assistance programs that are currently in other government departments, thereby assuming responsibility for U.S. development assistance across the board. The State Department would provide diplomatic support for this new agency, helping in the overall effort to reverse the process of state failure.

The DGS would be funded by shifting fiscal resources from the Department of Defense, which defines security almost exclusively in military terms. In effect, the DGS budget would become part of a new security budget. It would focus on the central sources of state failure by helping to stabilize population, restore environmental support systems, eradicate poverty, provide universal primary school education, and strengthen the rule of law through bolstering police forces, court systems, and, where needed, the military.

The DGS would make such issues as debt relief and market access an integral part of U.S. policy. It would also provide a forum to coordinate domestic and foreign policy, ensuring that domestic policies, such as cotton export subsidies or subsidies to convert grain into fuel for cars, do not weaken the economies of low-income countries or raise the price of food to unaffordable levels for the poor. A successful export-oriented farm sector often offers a path out of poverty for a poor country. The department would provide a focus for the United States to help lead a growing international effort to reverse the process of state failure. It would also encourage private investment in failing states by providing loan guarantees to spur development.

Thus far the process of state failure has largely been a one-way street, with hardly any countries reversing the process. Liberia is one of the few that have turned the tide. Following 14 years of cruel civil war that took 200,000 lives, *Foreign Policy*'s annual ranking of failing states showed Liberia ranking ninth in 2005. But things began to turn around that year with the election of Ellen Johnson-Sirleaf, a graduate of the Harvard Kennedy School and a former World Bank official, as president. A fierce effort to root out corruption along with a multinational U.N. Peacekeeping Force of up to 15,000 troops who maintain the peace, repair roads, schools, and hospitals, and train police have brought progress to this war-torn country. By 2010, Liberia had dropped to thirty-third on the list of failing states.

In *Prism* magazine, John W. Blaney, who served as U.S. Ambassador to Liberia from 2002 to 2005, describes how a dead state was gradually resuscitated and brought back to life. He writes about the exceptional role of a U.N. group that "led the way in developing and tailoring a disarmament, demobilization, reintegration, and rehabilitation program." He further notes that "we plotted out what should be done sequentially and simultaneously once the fighting stopped." Blaney concludes that there is no set formula for rebuilding a collapsed state—each situation is unique.

Collectively, the Plan B initiatives for education,

health, and family planning discussed in this chapter are estimated to cost another $75 billion a year. These cornerstones of human capital development and population stabilization will also help prevent state failure by alleviating the root social causes. Meanwhile, more effective responses to failing states can be paid for by redistributing donor countries' existing security budgets to reflect the twenty-first century threats they must address.

As Jeffrey Sachs regularly reminds us, for the first time in history we have the technological and financial resources to eradicate poverty. Investments in education, health, family planning, and school lunches are in a sense a humanitarian response to the plight of the world's poorest countries. But in the economically and politically integrated world of the twenty-first century, they are also an investment in our future.

Data, endnotes, and additional resources can be found on Earth Policy's Web site, at www.earth-policy.org.

12

Feeding Eight Billion

Prior to 1950, growth of the food supply came almost entirely from expanding cropland area. Then as frontiers disappeared and population growth accelerated after World War II, the focus quickly shifted to raising land productivity. In the most spectacular achievement in world agricultural history, farmers doubled the grain harvest between 1950 and 1973. Stated otherwise, growth in the grain harvest during this 23-year-span matched that of the preceding 11,000 years.

This was the golden age of world agriculture. Since then, growth in world food output has been gradually losing momentum as the backlog of unused agricultural technology dwindles, as soil erodes, as the area of cultivable land shrinks, and as irrigation water becomes scarce.

Gains in land productivity since 1950 have come primarily from three sources—the development of higher-yielding varieties, the growing use of fertilizer, and the spread of irrigation. The initial breakthrough in breeding higher-yielding varieties came when Japanese scientists succeeded in dwarfing both wheat and rice plants in the late nineteenth century. This decreased the share of photosynthate going into straw and increased that going into grain, making it possible to double yields.

With corn, now the world's leading grain crop, the early breakthrough came with hybridization in the United States. As a result of the dramatic advances associated with hybrid corn and the recent, much more modest gains associated with genetic modification, corn yields are still edging upward.

Most recently, Chinese scientists have developed commercially viable hybrid rice strains. While they have raised yields somewhat, the gains have been small compared with the earlier gains from dwarfing the rice plant.

As farmers attempted to remove nutrient constraints on crop yields, fertilizer use climbed from 14 million tons in 1950 to 163 million tons in 2009. In some countries, such as the United States, several in Western Europe, and Japan, fertilizer use has now leveled off or even declined substantially in recent decades. In China and India, both of which use more fertilizer than the United States does, usage may also decline as farmers use fertilizer more efficiently.

After several decades of rapid rise, however, it is now becoming more difficult to raise land productivity. From 1950 to 1990, world grainland productivity increased by 2.2 percent per year, but from 1990 until 2010 it went up only 1.2 percent annually.

There are distinct signs of yields leveling off in the higher-yield countries that are using all available technologies. With wheat, it is hard to get more than 8 tons per hectare. This is illustrated by the plateauing of wheat yields in France (Europe's largest wheat producer), Germany, the United Kingdom, and Egypt (Africa's leading wheat grower).

Japan, which led the world into the era of rising grain yields over a century ago, saw its rice yield plateau over the last decade or so as it approached 5 tons per hectare. Today yields in China are also leveling off as they reach the Japanese level.

Among the big three grains, corn is the only one where the yield is continuing a steady rise in high-yield countries. In the United States, which accounts for 40 percent of the world corn harvest, yields now exceed an astonishing 10 tons per hectare. Iowa, with its super-high corn yields, now produces more grain than Canada does.

Despite dramatic past leaps in grain yields, it is becoming more difficult to expand world food output for many reasons. Further gains in yields from plant breeding, even including genetic modification, do not come easily. Expanding the irrigated area is difficult. Returns on the use of additional fertilizer are diminishing in many countries.

In spite of the difficulties, some developing countries have dramatically boosted farm output. In India, after the monsoon failure of 1965 that required the import of a fifth of the U.S. wheat crop to avoid famine, a highly successful new agricultural strategy was adopted. It included replacing grain ceiling prices that catered to urban consumers with grain support prices to encourage farmers to invest in raising land productivity. The construction of fertilizer plants was moved from the public sector into the private sector, which could build them much faster. The high-yielding Mexican dwarf wheats, already tested in India, were introduced by the shipload for seed. These policy initiatives enabled India to double its wheat harvest in seven years. No major country before or since has managed to double the harvest of a staple food so quickly.

A similarly dramatic advance came in Malawi, a small country with low grain yields, after the drought of 2005 that left many hungry and some starving. In response, the government issued coupons good for 200 pounds of fertilizer to each farmer at well below the market price, along with free packets of improved seed corn, their staple food. Costing some $70 million per year and funded partly by outside donors, this fertilizer and seed subsidy program helped Malawi's farmers nearly double their corn harvest within two years, leading to an excess of grain. Fortunately this grain could be exported profitably to nearby Zimbabwe, which was experiencing acute grain shortages.

Some years earlier, Ethiopia, taking similar steps, also achieved a dramatic jump in production. But because there was no way to export the surplus, prices crashed—a major setback to the country's farmers. This experience underlines a major challenge to Africa's agricultural development, namely the lack of public infrastructure, such as roads to get fertilizer to farmers and their products to market.

In Africa's more arid countries, such as Chad, Mali, and Mauritania, there is not enough rainfall to raise yields dramatically. Modest yield gains are possible with improved agricultural practices, but in many of these low-rainfall countries there has not been a green revolution for the same reasons there has not been one in Australia—namely, low soil moisture and the associated limit on fertilizer use.

One encouraging practice to raise cropland productivity in semiarid Africa is the simultaneous planting of grain and nitrogen-fixing leguminous trees. At first the trees grow slowly, permitting the grain crop to mature and be harvested; then the saplings grow quickly to several feet in height, dropping leaves that provide nitrogen and organic matter, both sorely needed in African soils. The wood can then be cut and used for fuel. This simple, locally adapted technology, developed by scientists at the World Agroforestry Centre in Nairobi, has enabled farmers to double their grain yields within a matter of years as soil fertility builds.

The shrinking backlog of unused agricultural technology and the resulting loss of momentum in raising yields worldwide signals a need for fresh thinking on how to raise cropland productivity. One way is to breed crops that are more tolerant of drought and cold. U.S. corn breeders have developed corn strains that are more drought-tolerant, enabling corn production to move westward into Kansas, Nebraska, and South Dakota. For example, Kansas, the leading U.S. wheat-producing state, now produces more corn than wheat. Similarly, corn production is moving northward in North Dakota and Minnesota.

Another way to raise land productivity, where soil moisture permits, is to expand the land area that produces more than one crop per year. Indeed, the tripling of the world grain harvest from 1950 to 2000 was due in part to widespread increases in multiple cropping in Asia. Some of the more common combinations are wheat and corn in northern China, wheat and rice in northern India, and the double or triple cropping of rice in southern China and southern India.

The spread of corn-wheat double cropping on the North China Plain helped boost China's grain production to rival that of the United States. In northern India, the grain harvest 40 or so years ago was confined largely to wheat, but with the advent of the earlier maturing high-yielding wheats and rices, wheat could be harvested in time to plant rice. This combination is now widely used throughout the Punjab, Haryana, and parts of Uttar Pradesh.

Another often overlooked influence on productivity is land tenure. A survey by the Rural Development Institute revealed that farmers in China with documented land rights were twice as likely to make long-term investments in their land, such as adding greenhouses, orchards, or fishponds.

In summary, while grain production is falling in some countries, either because of unfolding water shortages or spreading soil erosion, the overwhelming majority of nations still have a substantial unrealized production potential. The challenge is for each country to fashion agricultural and economic policies to realize that potential. Countries like India in the late 1960s or Malawi in the last few years give a sense of how to exploit the possibilities for expanding food supplies.

With water shortages constraining food production growth, the world needs a campaign to raise water productivity similar to the one that nearly tripled land productivity over the last half-century. Data on the efficiency of surface water projects—that is, dams that deliver water to farmers through a network of canals—show that crops never use all the irrigation water simply because some evaporates, some percolates downward, and some runs off. Water policy analysts Sandra Postel and Amy Vickers found that "surface water irrigation efficiency ranges between 25 and 40 percent in India, Mexico, Pakistan, the Philippines, and Thailand; between 40 and 45 percent in Malaysia and Morocco; and between 50 and 60 percent in Israel, Japan, and Taiwan."

China's irrigation plan is to raise efficiency from 43 percent in 2000 to 55 percent in 2020. Key measures include raising the price of water, providing incentives for adopting more irrigation-efficient technologies, and developing the local institutions to manage this process.

Raising irrigation efficiency typically means shifting from the less-efficient flood or furrow systems to overhead sprinklers or to drip irrigation, the gold standard of irrigation efficiency. Switching from flood or furrow to low-pressure sprinkler systems reduces water use by an estimated 30 percent, while switching to drip irrigation typically cuts water use in half.

Drip irrigation also raises yields because it provides a steady supply of water with minimal losses to evaporation. In addition, it reduces the energy needed to pump water. Since drip systems are both labor-intensive and water-efficient, they are well suited to countries with a surplus of labor and a shortage of water. A few small countries—Cyprus, Israel, and Jordan—rely heavily on drip irrigation. This more-efficient technology is used on 1–3 percent of irrigated land in India and China and on roughly 4 percent in the United States.

In recent years, small-scale drip-irrigation systems—literally an elevated bucket with flexible plastic tubing to distribute the water—have been developed to irrigate small vegetable gardens with roughly 100 plants (covering 25 square meters). Somewhat larger systems using drums irrigate 125 square meters. In both cases, the containers are elevated slightly so that gravity distributes the water. Large-scale drip systems using plastic lines that can be moved easily are also becoming popular. These simple systems can pay for themselves in one year. By simultaneously reducing water costs and raising yields, they can dramatically raise incomes of smallholders.

Sandra Postel of the Global Water Policy Project estimates that drip technology has the potential to profitably irrigate 10 million hectares of India's cropland, nearly one tenth of the total. She sees a similar potential for China, which is now also expanding its drip irrigated area to save scarce water.

Institutional shifts—specifically, moving the responsibility for managing irrigation systems from government agencies to local water users associations—can facilitate the more efficient use of water. Farmers in many countries are organizing locally so they can assume this responsibility, and since they have an economic stake in good water management they tend to do a better job than a distant government agency. Mexico is a leader in developing water users associations. As of 2008, farmers associations managed more than 99 percent of the irrigated area held in public irrigation districts. One advantage of this shift is that the cost of maintaining the irrigation system is assumed locally, reducing the drain on the treasury.

Low water productivity is often the result of low water prices. In many countries, subsidies lead to irrationally low water prices, creating the impression that water is abundant when in fact it is scarce. As water becomes scarce, it needs to be priced accordingly.

A new mindset is needed, a new way of thinking about water use. For example, shifting to more water-efficient crops wherever possible boosts water productivity. Rice growing is being phased out around Beijing because rice is such a thirsty crop. Similarly, Egypt restricts rice production in favor of wheat. Any measures that raise crop yields on irrigated land also raise irrigation water productivity.

Bringing water use down to the sustainable yield of aquifers and rivers worldwide involves a wide range of measures not only in agriculture but throughout the economy. The more obvious steps, in addition to adopting more water-efficient irrigation practices, include using more water-efficient industrial processes. Recycling urban water supplies is another obvious step in countries facing acute water shortages. And because coal-fired power plants use so much water for cooling, shifting to wind farms eliminates a major drain on water supplies.

Another way to raise both land and water productivity is to produce animal protein more efficiently. With some 35 percent of the world grain harvest (760 million tons) used to produce animal protein, even a modest reduction in meat consumption or gain in efficiency can save a large quantity of grain.

World consumption of animal protein is everywhere on the rise. Meat consumption increased from 44 million tons in 1950 to 272 million tons in 2009, more than doubling annual consumption per person to nearly 90 pounds. The rise in consumption of milk and eggs is equally dramatic. Wherever incomes rise, so does meat consumption, reflecting a taste that apparently evolved over 4 million years of hunting and gathering.

As the oceanic fish catch and rangeland beef production have both leveled off, the world has shifted to grain-based production of animal protein to expand output. The efficiency with which various animals convert grain into protein varies widely. With cattle in feedlots, it takes roughly 7 pounds of grain to produce a 1-pound gain in live weight. For pork, the figure is over 3 pounds, for poultry it is just over 2, and for herbivorous species of farmed fish (such as carp, tilapia, and catfish), it is less than 2.

Global beef production, most of which comes from rangelands, grew less than 1 percent a year from 1990 to 2007 and has plateaued since. Pork production grew by 2 percent annually, and poultry by 4 percent. World pork production, half of it now in China, overtook beef production in 1979 and has continued to widen the lead since then. Poultry production eclipsed beef in 1995, moving into second place behind pork.

Fast-growing, grain-efficient fish farm output may also soon overtake beef production. In fact, aquaculture has been the fastest-growing source of animal protein since 1990, expanding from 13 million tons then to 52 million tons in 2008, or 8 percent a year.

Public attention has focused on aquacultural operations that are environmentally inefficient or disruptive, such as the farming of salmon, a carnivorous species that is typically fed fishmeal. But these operations account for less than one tenth of world fish farm output. Worldwide, aquaculture is dominated by herbivorous species—mainly carp in China and India, but also catfish in the United States and tilapia in several countries—and shellfish. This is where the great growth potential for efficient animal protein production lies.

China accounts for 62 percent of global fish farm output. Its output is dominated by finfish (mostly carp), which are grown in inland freshwater ponds, lakes, reservoirs, and rice paddies, and by shellfish (oysters and mussels), which are produced mostly in coastal regions. A multi-species system, using four types of carp that feed at different levels of the food chain, commonly boosts pond productivity over that of monocultures by at least half. China's fish farm output of 32 million tons is nearly triple U.S. beef output of 12 million tons.

Soybean meal is universally used in mixing feed for livestock, poultry, and fish. In 2010 the world's farmers produced 254 million tons of soybeans. Of this, an estimated 30 million tons were consumed directly as tofu or other meat substitutes. Some 220 million tons were crushed, yielding roughly 40 million tons of soybean oil and 170 million tons of highly valued high-protein meal.

Combining soybean meal with grain in a one-to-four ratio dramatically boosts the efficiency with which grain is converted into animal protein, sometimes nearly doubling it. Virtually the entire world, including the three largest meat producers—China, the United States, and Brazil—now relies heavily on soybean meal as a protein supplement in feed rations.

The heavy use of soybean meal to boost feed efficiency helps explain why the production of meat, milk, eggs, and farmed fish has climbed even though the 35 percent share of the world grain harvest used for feed has decreased slightly over the last 20 years. It also explains why world soybean production has multiplied 15-fold since 1950.

Mounting pressures on land and water resources have led to the evolution of some promising new animal protein production systems that are based on roughage rather than grain, such as milk production in India. Since 1970, India's milk production has increased fivefold, jumping from 21 million to 110 million tons in 2009. In 1997 India overtook the United States to become the world's leading producer of milk and other dairy products.

What is so remarkable is that India has built the world's largest dairy industry based not on grain but almost entirely on crop residues—wheat straw, rice straw, and corn stalks—and grass gathered from the roadside. The value of India's annual milk output now exceeds that of its rice harvest.

A second new protein production model, one that also relies on ruminants and roughage, has evolved in four provinces in eastern China—Hebei, Shangdong, Henan, and Anhui—where double cropping of winter wheat and corn is common. These provinces, dubbed the Beef Belt by Chinese officials, use crop residues to produce much of China's beef. This use of crop residues to produce milk in India and beef in China lets farmers reap a second harvest from the original grain crop, thus boosting both land and water productivity.

While people in developing countries are focusing on moving up the food chain, in many industrial countries there is a growing interest in fresh, locally produced foods. In the United States, this interest is driven both by concerns about the climate effects of transporting food from distant places and by the desire for fresh food that supermarkets with long supply chains can no longer deliver. This is reflected in the growth of both home gardens and local farmers' markets.

With the fast-growing local foods movement, diets are becoming more locally shaped and more seasonal. In the United States, this trend toward localization can be seen in the recent rise in farm numbers. Between the agricultural census of 2002 and that of 2007, the number of farms increased by nearly 80,000 to roughly 2.2 million. Many of the new farms, mostly smaller ones—and a growing share of them operated by women—cater to local markets. Some produce fresh fruits and vegetables exclusively for farmers' markets. Others, such as goat farms that produce milk, cheese, and meat, produce specialized products. With many specializing in organic food, the number of organic farms in the United States jumped from 12,000 in 2002 to 18,200 in 2007.

Many market outlets are opening up for local U.S. produce. Farmers' markets, where local farmers bring their produce for sale, increased from 1,755 in 1994 to over 6,100 in 2010, more than tripling over 16 years. These markets facilitate personal ties between producers and consumers that do not exist in the impersonal confines of a supermarket.

Many schools and universities are now making a point of buying local food because it is fresher, tastier, and more nutritious and it fits into new campus greening programs. Supermarkets are increasingly contracting seasonally with local farmers when produce is available. For example, in late 2010 Walmart announced a plan to buy more produce from local farmers for its stores. Upscale restaurants emphasize locally grown food on their menus. Some year-round food markets are evolving that supply only locally produced foods, including not only fresh produce but also meat, milk, cheese, eggs, and other farm products.

Home gardening was given a big boost in the spring of 2009 when First Lady Michelle Obama worked with children from a local school to dig up a piece of the White House lawn to start a vegetable garden. There was a precedent for this: Eleanor Roosevelt planted a White House victory garden during World War II. Her initiative encouraged millions of victory gardens, which eventually grew 40 percent of the nation's fresh produce.

Although it was much easier to expand home gardening during World War II, when the United States was much more rural, there is still a huge gardening potential—given that the grass lawns surrounding U.S. residences collectively cover some 18 million acres. Converting even a small share of this to fresh vegetables and fruit trees could make a meaningful contribution.

Many cities and small towns in the United States and England are creating community gardens that can be used by those who would otherwise not have access to land for gardening. Providing space for community gardens is now seen by many local governments as an essential service, like providing playgrounds or parks.

Urban gardens are gaining popularity throughout the world. A program organized by the U.N. Food and Agriculture Organization (FAO) to help cities in developing countries establish urban garden programs is being well received. In five cities in the Democratic Republic of the Congo, for example, it has helped 20,000 gardeners improve their vegetable growing operations. Market gardens in Kinshasa, the country's capital, produce an estimated 80,000 tons of vegetables per year, meeting 65 percent of the city's needs.

In the city of El Alto near La Paz, Bolivia, FAO supports a highly successful micro-garden program for low-income families. Using small, low-cost greenhouses covering about 50 square yards each, some 1,500 households grow fresh vegetables the year round. Some of the produce is consumed at home; some is sold at local markets.

School gardens are another welcome development. Children learn how food is produced, a skill often lacking in urban settings, and they may get their first taste of fresh salad greens or vine-ripened tomatoes. School gardens also provide fresh produce for school lunches. California, a leader in this area, has 6,000 school gardens.

Food from more-distant locations boosts carbon emissions while losing flavor and nutrition. A survey of food consumed in Iowa showed conventional produce traveled on average 1,500 miles, not including food imported from other countries. In contrast, locally grown produce traveled on average 56 miles—a huge difference in fuel use. And a study in Ontario, Canada, found that 58 imported foods traveled an average of 2,800 miles. In an oil-scarce world, consumers are worried about food security in a long-distance food economy.

The high prices of natural gas, which is used to make nitrogen fertilizer, and of phosphate, as reserves are depleted, suggest a much greater future emphasis on nutrient recycling—an area where small farmers producing for local markets have a distinct advantage over massive livestock and poultry feeding operations.

With food, as with energy, achieving security now depends on looking at the demand side of the equation as well as the supply side. We cannot rely solely on expanding production to reverse the deteriorating food situation of recent years. This is why a basic Plan B goal is to accelerate the shift to smaller families and halt the growth in world population at 8 billion by 2040.

An American living high on the food chain with a diet heavy in grain-intensive livestock products, including red meat, consumes twice as much grain as the average Italian and nearly four times as much as the average Indian. Adopting a Mediterranean diet can cut the grain footprint of Americans roughly in half, reducing carbon emissions accordingly.

Ensuring future food security was once the exclusive responsibility of the ministry of agriculture, but this is changing. The minister of agriculture alone, no matter how competent, can no longer be expected to secure food supplies. Indeed, efforts by the minister of health and family planning to lower human fertility may have a greater effect on future food security than efforts in the ministry of agriculture to raise land fertility.

Similarly, if ministries of energy cannot quickly cut carbon emissions, the world will face crop-shrinking heat waves that can massively and unpredictably reduce harvests. Saving the mountain glaciers whose ice melt irrigates much of the cropland in China and India during the dry season is the responsibility of the ministry of energy, not solely the ministry of agriculture.

If the ministries of forestry and agriculture cannot work together to restore tree cover and reduce floods and soil erosion, grain harvests will shrink not only in smaller countries like Haiti and Mongolia, as they are doing, but also in larger countries, such as Russia and Argentina—both wheat exporters.

And where water shortages restrict food output, it will be up to ministries of water resources to do everything possible to raise national water productivity. With water, as with energy, the principal potential now is in increasing efficiency, not expanding supply.

In a world where cropland is scarce and becoming more so, decisions made in ministries of transportation on whether to develop land-consuming, auto-centered transport systems or more-diversified systems that are much less land-intensive will directly affect world food security.

In the end, it is up to ministries of finance to reallocate resources in a way that recognizes the new threats to security posed by agriculture's deteriorating natural support systems, continuing population growth, human-driven climate change, and spreading water shortages. Since many ministries of government are involved, it is the head of state who must redefine security.

At the international level, we need to address the threat posed by growing climate volatility and the associated rise in food price volatility. The tripling of wheat, rice, corn, and soybean prices between 2007 and 2008 put enormous stresses on governments and low-income consumers. This price volatility also affects producers, since price uncertainty discourages investment by farmers.

In this unstable situation, a new mechanism to stabilize world grain prices is needed—in effect, a World Food Bank (WFB). This body would establish a support price and a ceiling price for wheat, rice, and corn. The WFB would buy grain when prices fell to the support level and return it to the market when prices reached the ceiling level, thus moderating price fluctuations in a way that would benefit both consumers and producers. The principal role of the WFB governing board, representing major exporting as well as importing countries, would be to establish the price levels for acquiring and releasing grain.

One simple way to improve food security is for the United States to eliminate the fuel ethanol subsidy and abolish the mandates that are driving the conversion of grain into fuel. This would help stabilize grain prices and buy some time in which to reverse the environmental and demographic trends that are undermining our future. It would also help relax the political tensions over food security that have emerged within importing countries.

And finally, we all have a role to play as individuals. Whether we decide to bike, bus, or drive to work will affect carbon emissions, climate change, and food security. The size of the car we drive to the supermarket and its effect on climate may indirectly affect the size of the bill at the supermarket checkout counter. At the family level, we need to hold the line at two children. And if we are living high on the food chain, we can eat less grain-intensive livestock products, improving our health while helping to stabilize climate. Food security is something in which we all have a stake—and a responsibility.

Data, endnotes, and additional resources can be found on Earth Policy's Web site, at www.earth-policy.org.

IV
WATCHING THE CLOCK

13

Saving Civilization

We need an economy for the twenty-first century, one that is in sync with the earth and its natural support systems, not one that is destroying them. The fossil fuel-based, automobile-centered, throwaway economy that evolved in western industrial societies is no longer a viable model—not for the countries that shaped it or for those that are emulating them. In short, we need to build a new economy, one powered with carbon-free sources of energy—wind, solar, and geothermal—one that has a diversified transport system and that reuses and recycles everything.

With Plan B we can change course and move onto a path of sustainable progress, but it will take a massive mobilization—at wartime speed. This plan, or something very similar to it, is our only hope.

The Plan B goals—stabilizing climate, stabilizing population, eradicating poverty, and restoring the economy's natural support systems—are mutually dependent. All are essential to feeding the world's people. It is unlikely that we can reach any one goal without reaching the others. Moving the global economy off the decline-and-collapse path depends on reaching all four goals.

The key to restructuring the economy is to get the market to tell the truth through full-cost pricing. For energy, this means putting a tax on carbon to reflect the full cost of burning fossil fuels and offsetting it with a reduction in the tax on income.

If the world is to move onto a sustainable path, we need economists who will calculate indirect costs and work with political leaders to incorporate them into market prices by restructuring taxes. This will require help from other disciplines, including ecology, meteorology, agronomy, hydrology, and demography. Full-cost pricing that will create an honest market is essential to building an economy that can sustain civilization and progress.

Some 2,500 economists, including nine Nobel Prize winners in economics, have endorsed the concept of tax shifts. Harvard economics professor and former chairman of George W. Bush's Council of Economic Advisors N. Gregory Mankiw wrote in *Fortune* magazine: "Cutting income taxes while increasing gasoline taxes would lead to more rapid economic growth, less traffic congestion, safer roads, and reduced risk of global warming—all without jeopardizing long-term fiscal solvency. This may be the closest thing to a free lunch that economics has to offer."

The failure of the market to reflect total costs can readily be seen with gasoline. The most detailed analysis available of gasoline's indirect costs is by the International Center for Technology Assessment. When added together, the many indirect costs to society—including climate change, oil industry tax breaks, military protection of the oil supply, oil industry subsidies, oil spills, and treatment of auto exhaust-related respiratory illnesses—total roughly $12 per gallon. If this external cost is added to the roughly $3 per gallon price of gasoline in the United States, gas would cost $15 a gallon. These are real costs. Someone bears them. If not us, our children.

If we can get the market to tell the truth, to have market prices that reflect the full cost of burning gasoline or coal, of deforestation, of overpumping aquifers, and of overfishing, then we can begin to create a rational economy. If we can create an honest market, then market forces will rapidly restructure the world energy economy. Phasing in full-cost pricing will quickly reduce oil and coal use. Suddenly wind, solar, and geothermal will become much cheaper than climate-disrupting fossil fuels.

We are economic decisionmakers, whether as corporate planners, government policymakers, investment bankers, or consumers. And we rely on the market for price signals to guide our behavior. But if the market gives us bad information, we make bad decisions, and that is exactly what has been happening.

We are being blindsided by a faulty accounting system, one that will lead to bankruptcy. As Øystein Dahle, former Vice President of Exxon for Norway and the North Sea, has observed: "Socialism collapsed because it did not allow the market to tell the economic truth. Capitalism may collapse because it does not allow the market to tell the ecological truth."

If we leave costs off the books, we risk bankruptcy. A decade ago, a phenomenally successful company named Enron was frequently on the covers of business magazines. It was, at one point, the seventh most valuable corporation in the United States. But when some investors began raising questions, Enron's books were audited by outside accountants. Their audit showed that Enron was bankrupt—worthless. Its stock that had been trading for over $90 a share was suddenly trading for pennies.

Enron had devised some ingenious techniques for leaving costs off the books. We are doing exactly the same thing, but on a global scale. If we continue with this practice, we too will face bankruptcy.

Another major flaw in our market economy is that it neither recognizes nor respects sustainable yield limits of natural systems. Consider, for example, the overpumping of aquifers. Once there is evidence that a water table is starting to fall, the first step should be to ban the drilling of new wells. If the water table continues to fall, then water should be priced at a rate that will reduce its use and stabilize the aquifer. Otherwise, there is a "race to the bottom" as wells are drilled ever deeper. When the aquifer is depleted, the water-based food bubble will burst, reducing harvests and driving up food prices.

Or consider deforestation. Proper incentives, such as a stumpage tax for each tree cut, would automatically shift harvesting from clearcutting to selective cutting, taking only the mature trees and protecting the forests.

Not only do we distort reality when we omit costs associated with burning fossil fuels from their prices, but governments actually subsidize their use, distorting reality even further. Worldwide, subsidies that encourage the production and use of fossil fuels add up to roughly $500 billion per year, compared with less than $50 billion for renewable energy, including wind, solar, and biofuels. In 2009, fossil fuel consumption subsidies included $147 billion for oil, $134 billion for natural gas, and $31 billion for coal. Governments are shelling out nearly $1.4 billion per day to further destabilize the earth's climate.

Iran, with a fossil fuel subsidy of $66 billion, is a leader in promoting gasoline use by pricing it at one fifth its market price. Following Iran on the list of countries that heavily subsidize fossil fuel use are Russia, Saudi Arabia, and India.

Carbon emissions could be cut in scores of countries by simply eliminating fossil fuel subsidies. Some countries are already doing this. Belgium, France, and Japan have phased out all subsidies for coal. Countries in the European Union may phase out coal subsidies entirely by 2014. President Obama has announced plans to start phasing out fossil fuel subsidies in 2011. As oil prices have climbed, a number of countries that held fuel prices well below world market prices have greatly reduced or eliminated their motor fuel subsidies because of the heavy fiscal cost. Among those reducing subsidies are China, Indonesia, and Nigeria.

A world facing economically disruptive climate change can no longer justify subsidies to expand the burning of coal and oil. A phaseout of oil consumption subsidies over the next decade would cut oil use by 4.7 million barrels per day in 2020. Eliminating all fossil fuel consumption subsidies by 2020 would cut global carbon emissions by nearly 6 percent and reduce government debt.

Shifting subsidies to the development of climate-benign energy sources such as wind, solar, and geothermal power will help stabilize the earth's climate. Moving subsidies from road construction to high-speed intercity rail construction could increase mobility, reduce travel costs, and lower carbon emissions.

Closely related to the need to restructure the economy is the need to redefine security. One of our legacies from the last century, which was dominated by two world wars and the cold war, is a sense of security that is defined almost exclusively in military terms. It so dominates Washington thinking that the U.S. foreign affairs budget of $701 billion in 2009 consisted of $661 billion for military purposes and $40 billion for foreign assistance and diplomatic programs.

Douglas Alexander, former U.K. Secretary of State for International Development, put it well in 2007: "In the 20th century a country's might was too often measured in what they could destroy. In the 21st century strength should be measured by what we can build together."

The good news is that in the United States the concept of redefining security is now permeating not only various independent think tanks but the Pentagon itself. A number of studies have looked at threats to U.S. interests posed by climate change, population growth, water shortages, and food shortages—key trends that contribute to political instability and lead to social collapse.

Although security is starting to be redefined in a conceptual sense, we have not redefined it in fiscal terms. The United States still has a huge military budget, committed to developing and manufacturing technologically sophisticated and costly weapon systems. Since there is no other heavily armed superpower, the United States is essentially in an arms race with itself. What if the next war is fought in cyberspace or with terrorist insurgents? Vast investments in conventional weapons systems will be of limited use.

Given the enormity of the antiquated military budget, no one can argue that we do not have the resources to rescue civilization. The far-flung U.S. military establishment, including hundreds of military bases scattered around the world, will not save civilization. It belongs to another era. We can most effectively achieve our security goals by helping to expand food production, by filling the family planning gap, by building wind farms and solar power plants, and by building schools and clinics.

During the years when governments and the media were focused on preparing for the 2009 Copenhagen climate negotiations, a powerful movement opposing the construction of new coal-fired power plants was emerging in the United States, largely below the radar screen. The principal reason that environmental groups, both national and local, are opposing coal plants is that they are the primary driver of climate change. In addition, emissions from coal plants are responsible for 13,200 U.S. deaths annually—a number that dwarfs the U.S. lives lost in Iraq and Afghanistan combined.

Over the last few years the U.S. coal industry has suffered one setback after another. What began as a few local ripples of resistance to coal-fired power quickly evolved into a national tidal wave of grassroots opposition from environmental, health, farm, and community organizations. Despite a heavily funded industry campaign to promote "clean coal," the American public is turning against coal. In a national poll that asked which electricity source people would prefer, only 3 percent chose coal. The Sierra Club, which has kept a tally of proposed coal-fired power plants and their fates since 2000, reports that 139 plants in the United States have been defeated or abandoned.

An early turning point in the coal war came in June 2007, when Florida's Public Service Commission refused to license a huge $5.7-billion, 1,960-megawatt coal plant because the utility proposing it could not prove that building the plant would be cheaper than investing in conservation, efficiency, or renewable energy sources. This point, frequently made by lawyers from Earthjustice, a nonprofit environmental legal group, combined with widely expressed public opposition to any more coal-fired power plants in Florida, led to the quiet withdrawal of four other coal plant proposals in the state.

Coal's future also suffered as Wall Street, pressured by the Rainforest Action Network, turned its back on the industry. In early February 2008, investment banks Morgan Stanley, Citi, and J.P. Morgan Chase announced that any future lending for coal-fired power would be contingent on the utilities demonstrating that the plants would be economically viable with the higher costs associated with future federal restrictions on carbon emissions. Later that month, Bank of America announced it would follow suit.

One of the unresolved questions haunting the coal sector is what to do with the coal ash—the remnant of burning coal—that is accumulating in 194 landfills and 161 holding ponds in 47 states. This ash is not an easy material to dispose of since it is laced with arsenic, lead, mercury, and other toxic materials. The industry's dirty secret came into full public view just before Christmas 2008 when a Tennessee Valley Authority (TVA) coal ash pond containment wall in eastern Tennessee collapsed, releasing a billion gallons of toxic brew.

Surprising through it may seem, the industry does not have a plan for safely disposing of the 130 million tons of ash produced each year, enough to fill 1 million railroad cars. The spill of toxic coal ash in Tennessee, which is costing the TVA $1.2 billion to clean up, drove another nail into the lid of the coal industry coffin.

An August 2010 joint study by the Environmental Integrity Project, Earthjustice, and the Sierra Club reported that 39 coal ash dump sites in 21 states have contaminated local drinking water or surface water with arsenic, lead, and other heavy metals at levels that exceed federal safe drinking water standards. This is in addition to 98 coal ash sites that are polluting local water supplies that were already identified by the U.S. Environmental Protection Agency (EPA). In response to these and other threats, new regulations are in the making to require an upgrade of the management of coal ash storage facilities so as to avoid contaminating local groundwater supplies. In addition, EPA is issuing more stringent regulations on coal plant emissions, including sulfur dioxide and nitrogen oxides. The goal is to reduce chronic respiratory illnesses, such as asthma in children, and the deaths caused by coal-fired power plant emissions.

Another coal industry practice, the blasting off of mountain tops with explosives to get at coal seams, is under fire. In August 2010, the Rainforest Action Network announced that several leading U.S. investment banks, including Bank of America, J.P. Morgan, Citi, Morgan Stanley, and Wells Fargo, had ceased lending to companies involved in mountaintop removal coal mining. Massey Energy, a large coal mining company notorious for its violations of environmental and safety regulations and the owner of the West Virginia mine where 29 miners died in 2010, lost all funding from three of the banks.

More and more utilities are beginning to recognize that coal is not a viable long-term option. TVA, for example, announced in August 2010 that it was planning to close 9 of its 59 coal-generating units. Duke Energy, another major southeastern utility, followed with an announcement that it was considering the closure of seven coal-fired units in North and South Carolina alone. Progress Energy, also in the Carolinas, is planning to close 11 units at four sites. In Pennsylvania, Exelon Power is preparing to close four coal units at two sites. And Xcel Energy, the dominant utility in Colorado, announced it was closing seven coal units.

These five are examples of a growing number of U.S. utilities that are closing coal-fired power plants, replacing them with natural gas, wind, solar, biomass, and efficiency gains. In an analysis of the future of coal, Wood Mackenzie, a leading energy consulting and research firm, sees these closings as a harbinger of things to come for the coal industry.

The chairman of the powerful U.S. Federal Energy Regulatory Commission, Jon Wellinghoff, observed in early 2009 that the United States may no longer need any additional coal plants. Regulators, investment banks, and political leaders are now beginning to see what has been obvious for some time to climate scientists such as James Hansen: that it makes no sense to build coal-fired power plants only to have to bulldoze them in a few years.

Given the huge potential for reducing electricity use in the United States, closing coal plants may be much easier than it appears. If the efficiency level of the other 49 states were raised to that of New York, the most energy-efficient state, the energy saved would be sufficient to close 80 percent of the country's coal-fired power plants. The remaining plants could be shut down by turning to renewable energy—wind farms, solar thermal power plants, solar cells, and geothermal power and heat.

As noted earlier, the U.S. transition from coal to renewables is under way. Between 2007 and 2010, U.S. coal use dropped 8 percent. During the same period, and despite the recession, 300 new wind farms came online, adding some 21,000 megawatts of wind-generating capacity.

The bottom line is that the United States currently has, in effect, a near de facto moratorium on the licensing of new coal-fired power plants. Several environmental groups, including the Sierra Club and Greenpeace, are now starting to focus on closing existing coal plants. The movement is also going international, as campaigns are now under way in several countries to prevent the construction of new coal plants and to close existing ones.

With the likelihood that few, if any, new coal-fired power plants will be approved in the United States, this moratorium sends a message to the world. Denmark and New Zealand have already banned new coal-fired power plants. Hungary is on the verge of closing its one remaining coal plant. Ontario Province, where 39 percent of Canadians live, plans to phase out coal entirely by 2014. Scotland announced in September 2010 that it plans to get 80 percent of its electricity from renewables by 2020 and 100 percent by 2025, backing out coal entirely. Other countries are likely to join this effort to cut carbon emissions. Even China, which was building one new coal plant a week, is surging ahead with renewable energy and now leads the world in new wind farm installations. These and other developments suggest that the Plan B goal of cutting carbon emissions 80 percent by 2020 may be much more attainable than many would have thought a few years ago.

The restructuring of the energy economy will not only dramatically drop carbon emissions, helping to stabilize climate, it will also eliminate much of the air pollution that we know today. The idea of a pollution-free environment is difficult for us even to imagine, simply because none of us has ever known an energy economy that was not highly polluting. Working in coal mines will be history. Black lung disease will eventually disappear. So too will "code red" alerts warning us to avoid strenuous exercise because of dangerous levels of air pollution.

And, finally, in contrast to investments in oil fields and coal mines, where depletion and abandonment are inevitable, the new energy sources are inexhaustible. While wind turbines, solar cells, and solar thermal systems will all need repair and occasional replacement, investing in these new energy sources means investing in energy systems that can last forever. These wells will not go dry.

Although some of the prospects look good for moving away from coal, timing is key. Can we close coal-fired power plants fast enough to save the Greenland ice sheet? To me, saving Greenland is both a metaphor and a precondition for saving civilization. If its ice sheet melts, sea level will rise 23 feet. Hundreds of coastal cities will be abandoned. The rice-growing river deltas of Asia will be under water. And there will be hundreds of millions of rising-sea refugees. The word that comes to mind is chaos. If we cannot mobilize to save the Greenland ice sheet, we probably cannot save civilization as we know it.

Similarly, can we eradicate poverty and fill the family planning gap fast enough to help countries escape the demographic trap? Can we halt the growth in the number of failing states before our global civilization begins to unravel?

The overarching question is, Can we change fast enough? When thinking about the enormous need for social change as we attempt to move the world economy onto a sustainable path, I find it useful to look at three models of social change. One is the Pearl Harbor model, where a dramatic event fundamentally changed how Americans thought and behaved. The second model is one where a society reaches a tipping point on a particular issue often after an extended period of gradual change in thinking and attitudes. This I call the Berlin Wall model. The third is the sandwich model of social change, where there is a dedicated grassroots movement pushing for change that is strongly supported by political leadership.

The surprise Japanese attack on Pearl Harbor on December 7, 1941, was a dramatic wakeup call. It totally changed how Americans thought about the war. If the American people had been asked on December 6th whether the country should enter World War II, probably 95 percent would have said no. By Monday morning, December 8th, 95 percent would likely have said yes.

When scientists are asked to identify a possible "Pearl Harbor" scenario on the climate front, they frequently point to the possible breakup of the West Antarctic ice sheet. Sizable blocks of it have been breaking off for more than a decade already, but far larger blocks could break off, sliding into the ocean. Sea level could rise a frightening 2 or 3 feet within a matter of years. Unfortunately, if we reach this point it may be too late to cut carbon emissions fast enough to save the remainder of the West Antarctic ice sheet. By then we might be over the edge.

The Berlin Wall model is of interest because the wall's dismantling in November 1989 was a visual manifestation of a much more fundamental social change. At some point, Eastern Europeans, buoyed by changes in Moscow, had rejected the great "socialist experiment" with its one-party political system and centrally planned economy. Although it was not anticipated, Eastern Europe had an essentially bloodless revolution, one that changed the form of government in every country in the region. It had reached a tipping point.

Many social changes occur when societies reach tipping points or cross key thresholds. Once that happens, change comes rapidly and often unpredictably. One of the best known U.S. tipping points is the growing opposition to smoking that took place during the last half of the twentieth century. This movement was fueled by a steady flow of information on the health-damaging effects of smoking, a process that began with the Surgeon General's first report in 1964 on smoking and health. The tipping point came when this information flow finally overcame the heavily funded disinformation campaign of the tobacco industry.

Although many Americans are confused by the disinformation campaign on climate change, which is funded by the oil and coal industries, there are signs that the United States may be moving toward a tipping point on climate, much as it did on tobacco in the 1990s. The oil and coal companies are using some of the same disinformation tactics that the tobacco industry used in trying to convince the public that there was no link between smoking and health.

The sandwich model of social change is in many ways the most attractive one, largely because of its potential for rapid change, as with the U.S. civil rights movement in the 1960s. Strong steps by EPA to enforce existing laws that limit toxic pollutants from coal-fired power plants, for instance, are making coal much less attractive. So too do the regulations on managing coal ash storage and rulings against mountaintop removal. This, combined with the powerful grassroots campaign forcing utilities to seek the least cost option, is spelling the end of coal.

Of the three models of social change, relying on the Pearl Harbor model for change is by far the riskiest, because by the time a society-changing catastrophic event occurs for climate change, it may be too late. The Berlin Wall model works, despite the lack of government support, but it does take time. The ideal situation for rapid, historic progress occurs when mounting grassroots pressure for change merges with a national leadership that is similarly committed.

Whenever I begin to feel overwhelmed by the scale and urgency of the changes we need to make, I reread the economic history of U.S. involvement in World War II because it is such an inspiring study in rapid mobilization. Initially, the United States resisted involvement in the war and responded only after it was directly attacked at Pearl Harbor. But respond it did. After an all-out commitment, the U.S. engagement helped turn the tide of war, leading the Allied Forces to victory within three-and-a-half years.

In his State of the Union address on January 6, 1942, one month after the bombing of Pearl Harbor, President Franklin D. Roosevelt announced the country's arms production goals. The United States, he said, was planning to produce 45,000 tanks, 60,000 planes, and several thousand ships. He added, "Let no man say it cannot be done."

No one had ever seen such huge arms production numbers. Public skepticism abounded. But Roosevelt and his colleagues realized that the world's largest concentration of industrial power was in the U.S. automobile industry. Even during the Depression, the United States was producing 3 million or more cars a year.

After his State of the Union address, Roosevelt met with auto industry leaders, indicating that the country would rely heavily on them to reach these arms production goals. Initially they expected to continue making cars and simply add on the production of armaments. What they did not yet know was that the sale of new cars would soon be banned. From early February 1942 through the end of 1944, nearly three years, essentially no cars were produced in the United States.

In addition to a ban on the sale of new cars, residential and highway construction was halted, and driving for pleasure was banned. Suddenly people were recycling and planting victory gardens. Strategic goods—including tires, gasoline, fuel oil, and sugar—were rationed beginning in 1942. Yet 1942 witnessed the greatest expansion of industrial output in the nation's history—all for military use. Wartime aircraft needs were enormous. They included not only fighters, bombers, and reconnaissance planes, but also the troop and cargo transports needed to fight a war on distant fronts. From the beginning of 1942 through 1944, the United States far exceeded the initial goal of 60,000 planes, turning out a staggering 229,600 aircraft, a fleet so vast it is hard even today to visualize it. Equally impressive, by the end of the war more than 5,000 ships were added to the 1,000 or so that made up the American Merchant Fleet in 1939.

In her book *No Ordinary Time*, Doris Kearns Goodwin describes how various firms converted. A sparkplug factory switched to the production of machine guns. A manufacturer of stoves produced lifeboats. A merry-go-round factory made gun mounts; a toy company turned out compasses; a corset manufacturer produced grenade belts; and a pinball machine plant made armor-piercing shells.

In retrospect, the speed of this conversion from a peacetime to a wartime economy is stunning. The harnessing of U.S. industrial power tipped the scales decisively toward the Allied Forces, reversing the tide of war. Germany and Japan, already fully extended, could not counter this effort. British Prime Minister Winston Churchill often quoted his foreign secretary, Sir Edward Grey: "The United States is like a giant boiler. Once the fire is lighted under it, there is no limit to the power it can generate."

The point is that it did not take decades to restructure the U.S. industrial economy. It did not take years. It was done in a matter of months. If we could restructure the

U.S. industrial economy in months, then we can restructure the world energy economy during this decade.

With numerous U.S. automobile assembly lines currently idled, it would be a relatively simple matter to retool some of them to produce wind turbines, as the Ford Motor Company did in World War II with B-24 bombers, helping the world to quickly harness its vast wind energy resources. This would help the world see that the economy can be restructured quickly, profitably, and in a way that enhances global security.

The world now has the technologies and financial resources to stabilize climate, eradicate poverty, stabilize population, restore the economy's natural support systems, and, above all, restore hope. The United States, the wealthiest society that has ever existed, has the resources and leadership to lead this effort.

We can calculate roughly the costs of the changes needed to move our twenty-first century civilization off the decline-and-collapse path and onto a path that will sustain civilization. What we cannot calculate is the cost of not adopting Plan B. How do you put a price tag on social collapse and the massive die-off that it invariably brings?

As noted in earlier chapters, the external funding needed to eradicate poverty and stabilize population requires an additional expenditure of $75 billion per year. A poverty eradication effort that is not accompanied by an earth restoration effort is doomed to fail. Protecting topsoil, reforesting the earth, restoring oceanic fisheries, and other needed measures will cost an estimated $110 billion in additional expenditures per year. Combining both social goals and earth restoration goals into a Plan B budget yields an additional annual expenditure of $185 billion. (See Table 13–1.) This is the new defense budget, the one that addresses the most serious threats to both national and global security. It is equal to 12 percent of

Table 13–1. *Plan B Budget: Additional Annual Expenditures Needed to Meet Social Goals and Restore the Earth*

Goal	Funding (billion dollars)
Basic Social Goals	
Universal primary education	10
Eradication of adult illiteracy	4
School lunch programs	3
Aid to women, infants, preschool children	4
Reproductive health and family planning	21
Universal basic health care	33
Total	75
Earth Restoration Goals	
Planting trees	23
Protecting topsoil on cropland	24
Restoring rangelands	9
Restoring fisheries	13
Stabilizing water tables	10
Protecting biological diversity	31
Total	110
Grand Total	185
U.S. Military Budget	661
Plan B budget as share of this	28%
World Military Budget	1,522
Plan B budget as share of this	12%

Source: Military from SIPRI; other data at www.earth-policy.org.

global military expenditures and 28 percent of U.S. military expenditures.

Unfortunately, the United States continues to focus its fiscal resources on building an ever-stronger military, largely ignoring the threats posed by continuing environmental deterioration, poverty, and population growth. Its 2009 military expenditures accounted for 43 percent of the global total of $1,522 billion. Other leading spenders included China ($100 billion), France ($64 billion), the United Kingdom ($58 billion), and Russia ($53 billion).

For less than $200 billion of additional funding per year worldwide, we can get rid of hunger, illiteracy, disease, and poverty, and we can restore the earth's soils, forests, and fisheries. We can build a global community where the basic needs of all people are satisfied—a world that will allow us to think of ourselves as civilized.

As a general matter, the benchmark of political leadership will be whether leaders succeed in shifting taxes from work to environmentally destructive activities. It is tax shifting, not additional appropriations, that is the key to restructuring the energy economy in order to stabilize climate.

Just as the forces of decline can reinforce each other, so too can the forces of progress. For example, efficiency gains that lower oil dependence also reduce carbon emissions and air pollution. Eradicating poverty helps stabilize population. Reforestation sequesters carbon, increases aquifer recharge, and reduces soil erosion. Once we get enough trends headed in the right direction, they will reinforce each other.

One of the questions I hear most frequently is, What can I do? People often expect me to suggest lifestyle changes, such as recycling newspapers or changing light bulbs. These are essential, but they are not nearly enough. Restructuring the global economy means becoming politically active, working for the needed changes, as the grassroots campaign against coal-fired power plants is doing. Saving civilization is not a spectator sport.

Inform yourself. Read about the issues. Share this book with friends. Pick an issue that's meaningful to you, such as tax restructuring to create an honest market, phasing out coal-fired power plants, or developing a world class-recycling system in your community. Or join a group that is working to provide family planning services to the 215 million women who want to plan their families but lack the means to do so. You might want to organize a small group of like-minded individuals to work on an issue that is of mutual concern. You can begin by talking with others to help select an issue to work on.

Once your group is informed and has a clearly defined goal, ask to meet with your elected representatives on the city council or the state or national legislature. Write or e-mail your elected representatives about the need to restructure taxes and eliminate fossil fuel subsidies. Remind them that leaving environmental costs off the books may offer a sense of prosperity in the short run, but it leads to collapse in the long run.

During World War II, the military draft asked millions of young men to risk the ultimate sacrifice. But we are called on only to be politically active and to make lifestyle changes. During World War II, President Roosevelt frequently asked Americans to adjust their lifestyles and Americans responded, working together for a common goal. What contributions can we each make today, in time, money, or reduced consumption, to help save civilization?

The choice is ours—yours and mine. We can stay with business as usual and preside over an economy that continues to destroy its natural support systems until it destroys itself, or we can be the generation that changes direction, moving the world onto a path of sustained progress. The choice will be made by our generation, but it will affect life on earth for all generations to come.

Data, endnotes, and additional resources can be found on Earth Policy's Web site, at www.earth-policy.org.

Additional Resources

More information on the topics covered in World on the Edge *can be found in the references listed here. The full text of the book, along with extensive endnotes, datasets, and new releases, is available on the Earth Policy Institute Web site at www.earth-policy.org.*

Chapter 1

Herman E. Daly, "Economics in a Full World," *Scientific American*, vol. 293, no. 3 (September 2005), pp. 100–07.

Jared Diamond, *Collapse: How Societies Choose to Fail or Succeed* (New York: Penguin Group, 2005).

Global Footprint Network, WWF, and Zoological Society of London, *Living Planet Report 2010* (Gland, Switzerland: WWF, October 2010).

Mathis Wackernagel et al., "Tracking the Ecological Overshoot of the Human Economy," *Proceedings of the National Academy of Sciences*, vol. 99, no. 14 (9 July 2002), pp. 9,266–71.

Ronald A. Wright, *A Short History of Progress* (New York: Carroll and Graf Publishers, 2005).

Chapter 2

John Briscoe, *India's Water Economy: Bracing for a Turbulent Future* (New Delhi: World Bank, 2005).

Sanjay Pahuja et al., *Deep Wells and Prudence: Towards Pragmatic Action for Addressing Ground-water Over-exploitation in India* (Washington, DC: World Bank, January 2010).

Sandra Postel, *Pillar of Sand* (New York: W. W. Norton & Company, 1999).

Tushaar Shah, *Taming the Anarchy: Groundwater Governance in South Asia* (Washington, DC: RFF Press, 2009).

U.N. Food and Agriculture Organization, "AQUASTAT: Countries and Regions," at www.fao.org/nr/water/aquastat/countries/index.stm.

Chapter 3

David R. Montgomery, *Dirt: The Erosion of Civilizations* (Berkeley, CA: University of California Press, 2007).

NASA Earth Observatory, at earthobservatory.nasa.gov.

U.N. Convention to Combat Desertification, at www.unccd.int.

Chapter 4

Joseph Romm, *Climate Progress*, blog at www.climateprogress.org.

James Hansen, "How Warm Was This Summer?" 1 October 2010, at www.columbia.edu/~jeh1/mailings.

David B. Lobell and Christopher B. Field, "Global Scale Climate-Crop Yield Relationships and the Impacts of Recent Warming," *Environmental Research Letters*, vol. 2, no. 1 (16 March 2007).

National Snow and Ice Data Center, at nsidc.org.

U.N. Environment Programme, *Global Outlook for Ice and Snow* (Nairobi, Kenya: 2007).

Chapter 5

GRAIN, *Food Crisis and the Global Land Grab*, blog and news archive at farmlandgrab.org.

Joachim von Braun and Ruth Meinzen-Dick, "*Land Grabbing*" *by Foreign Investors in Developing Countries*, Policy Brief No. 13 (Washington, DC: International Food Policy Research Institute, April 2009).

U.N. Food and Agriculture Organization, *The State of Food Insecurity in the World 2010* (Rome: 2010).

World Bank, *Rising Global Interest in Farmland: Can It Yield Sustainable and Equitable Benefits?* (Washington, DC: September 2010).

Chapter 6

Environmental Justice Foundation, *No Place Like Home: Where Next for Climate Refugees?* (London: 2009).

Gordon McGranahan et al., "The Rising Tide: Assessing the Risks of Climate Change and Human Settlements in Low Elevation Coastal Zones," *Environment and Urbanization*, vol. 18, no. 1 (April 2007).

Koko Warner et al., *In Search of Shelter: Mapping the*

Effects of Climate Change on Human Migration and Displacement (Atlanta, GA: CARE International, 2009).

Chapter 7

Pauline H. Baker, "Forging a U.S. Policy Toward Fragile States," *Prism*, vol. 1, no. 2 (March 2010).

Fund for Peace and *Foreign Policy*, "The Failed States Index," *Foreign Policy*, July/August various years, with full index on-line at www.fundforpeace.org.

Political Instability Task Force, at globalpolicy.gmu.edu/pitf.

Susan E. Rice and Stewart Patrick, *Index of State Weakness in the Developing World* (Washington, DC: Brookings, 2008).

Chapter 8

Complete Streets Coalition, at www.completestreets.org.

Institute for Transportation and Development Policy, at www.itdp.org.

International Energy Agency, *Energy Technology Perspectives 2010* (Paris: 2010).

McKinsey & Co., *Pathways to a Low-Carbon Economy* (New York: 2009).

Natalie Mims, Mathias Bell, and Stephen Doig, *Assessing the Electric Productivity Gap and the U.S. Efficiency Opportunity* (Snowmass, CO: Rocky Mountain Institute, January 2009).

U.S. Green Building Council, at www.usgbc.org.

Chapter 9

Alison Holm, Leslie Blodgett, Dan Jennejohn, and Karl Gawell, *Geothermal Energy International Market Update* (Washington, DC: Geothermal Energy Association, May 2010).

European Photovoltaic Industry Association, *Global Market Outlook for Photovoltaics Until 2014* (Brussels: May 2010).

Global Wind Energy Council, *Global Wind 2009 Report* (Brussels: 2010).

Renewable Energy Policy Network for the 21st Century, *Renewables 2010 Global Status Report* (Paris: 2010).

Xi Lu, Michael B. McElroy, and Juha Kiviluoma, "Global Potential for Wind-Generated Electricity," *Proceedings of the National Academy of Sciences*, vol. 106, no. 27 (7 July 2009), pp. 10,933–38.

Chapter 10

Andrew Balmford et al., "The Worldwide Costs of Marine Protected Areas," *Proceedings of the National Academy of Sciences*, vol. 101, no. 26 (29 June 2004), pp. 9,694–97.

Johan Eliasch, *Climate Change: Financing Global Forests* (London: Her Majesty's Stationery Office, 2008).

U.N. Environment Programme, Billion Tree Campaign, at www.unep.org/billiontreecampaign.

U.N. Food and Agriculture Organization, *Forest Resources Assessment 2010* (Rome: 2010), with additional information at www.fao.org/forestry.

Chapter 11

Alex Duncan, Gareth Williams, and Juana de Catheu, *Monitoring the Principles for Good International Engagement in Fragile States and Situations* (Paris: Organisation for Economic Co-operation and Development, 2010).

Jeffrey D. Sachs and the Commission on Macroeconomics and Health, *Macroeconomics and Health: Investing in Health for Economic Development* (Geneva: World Health Organization, 2001), at www.paho.org/English/DPM/SHD/HP/Sachs.pdf.

Susheela Singh et al., *Adding It Up: The Costs and Benefits of Investing in Family Planning and Maternal and Newborn Health* (New York: Guttmacher Institute and United Nations Population Fund, 2009).

U.N. Department of Social and Economic Affairs, *Millennium Development Goals Report 2010* (New York: June 2010), with more on the Millennium Development Goals at www.un.org/millenniumgoals.

UNESCO, *Education for All Global Monitoring Report 2010: Reaching the Marginalized* (Paris: 2010).

World Bank and International Monetary Fund, *Global Monitoring Report 2010: The MDGs after the Crisis* (Washington, DC: 2010).

Chapter 12

Consultative Group on International Agricultural Research, at www.cgiar.org.

M. Herrero et al., "Smart Investments in Sustainable Food Production: Revisiting Mixed Crop-Livestock

Systems," *Science*, vol. 327, no. 5967 (12 February 2010), pp. 822–25.

International Water Management Institute, at www.iwmi.cgiar.org.

National Research Council, *Toward Sustainable Agricultural Systems in the 21st Century* (Washington, DC: National Academies Press, 2010).

Sandra Postel and Amy Vickers, "Boosting Water Productivity," in Worldwatch Institute, *State of the World 2004* (New York: W. W. Norton & Company, 2004).

U.N. Food and Agriculture Organization, *Growing Greener Cities* (Rome: 2010).

U.N. Food and Agriculture Organization, *The State of World Fisheries and Aquaculture* (Rome: various years).

Chapter 13

Carbon Tax Center, at www.carbontax.org.

The CNA Corporation, *National Security and the Threat of Climate Change* (Alexandria, VA: 2007), at www.cna.org/reports/climate.

International Center for Technology Assessment, *The Real Cost of Gasoline: An Analysis of the Hidden External Costs Consumers Pay to Fuel Their Automobiles* (Washington, DC: 1998), with updates from *Gasoline Cost Externalities* (Washington, DC: 2004 and 2005).

Ted Nace, *Climate Hope* (San Francisco: Coal Swarm, 2010), with additional information at the Coal Swarm Web site, at www.sourcewatch.org/ index.php?title= CoalSwarm.

Sierra Club, *Stopping the Coal Rush*, at www.sierra club.org/environmentallaw/coal/plantlist.aspx.

U.S. Department of Defense, *Quadrennial Defense Review Report* (Washington, DC: February 2010).

Francis Walton, *Miracle of World War II: How American Industry Made Victory Possible* (New York: Macmillan, 1956).

Index

Acknowledgments

As I have noted before, if it takes a village to raise a child, then it takes the entire world to write a book of this scope. We draw on the work of thousands of scientists and research teams in many fields throughout the world. The process ends with the teams who translate the book into other languages.

In between, and most important, are the research team, reviewers, and staff at the Earth Policy Institute (EPI). The research team is led by Janet Larsen, our Director of Research. Janet is also my alter ego, my best critic, and a sounding board for new ideas. In researching for this book, the team combed through thousands of research reports, articles, and books—gathering, organizing, and analyzing information.

J. Matthew Roney and Alexandra Giese anchored a heroic research effort, feeding me a constant stream of new and valuable data. Amy Heinzerling played a key role until she left for graduate work. Interns Lauren Kubiak and Brigid Fitzgerald Reading helped with data gathering, fact checking, and reviewing. I am deeply grateful to each of them for their unflagging enthusiasm and dedication.

Some authors write, but this one dictates. Many thanks to Kristina Taylor who transcribed the many drafts and who also anchors the social networking part of EPI.

Reah Janise Kauffman, our Vice President, not only manages the Institute, thus enabling me to concentrate on research, she also directs EPI's outreach effort. This includes, among other things, coordinating our worldwide network of publishers, organizing book tours, and working with the media. (The ice-calving photo on the jacket was her idea.) Reah Janise's productivity and versatility are keys to the Institute's success. Her value to me is evidenced in our 24 years of working together as a team.

Millicent Johnson, our Manager of Publications Sales, handles our publications department and serves as our office quartermaster and librarian. Millicent cheerfully handles the thousands of book orders and takes pride in her one-day turnaround policy.

A number of reviewers helped shape the final product. My colleagues at EPI reviewed several drafts, providing insightful comments and suggestions. Reviewers from outside the Institute include Doug and Debra Baker, who, with their wide-ranging scientific knowledge from physics to meteorology, provided detailed comments. Also providing useful feedback were Peter Goldmark, for many years publisher of the *International Herald Tribune*; Edwin (Toby) Clark, formerly deputy administrator at the U.S. Environmental Protection Agency; William Mansfield, a member of the EPI Board and former Deputy Director of the United Nations Environment Programme; Maureen Kuwano Hinkle, with 26 years of experience working on agricultural issues with the Environmental Defense Fund and the Audubon Society; Frances Moore, a former EPI researcher now in graduate school; and Jessica Robbins, a former intern.

My thanks also to individuals who were particularly helpful in providing specific information: Upali Amarasinghe, Mathias Bell, Amos Bromhead, Colin J. Campbell, Martha M. Campbell, Jim Carle, Shaohua Chen, Robert W. Corell, Alberto Del Lungo, Rolf Derpsch, Junko Edahiro, Mark Ellis, David Fridley, Reed Funk, Nathan Glasgow, Monique Hanis, Bill Heenan, Ryde James, Michael Kintner-Meyer, Doug Koplow, Felix Kramer, Kathleen Krust, Rattan Lal, Li Junfeng, Eric Martinot, Heitor Matallo, Hirofumi Muraoka, Margaret Newman, Hassan Partow, John Pucher, Richard Register, William Ryerson, Richard Schimpf, Stefanie Seskin, John E. Sheehy, Ashbindu Singh, Swati Singh, Kara Slack, J. Joseph Speidel, Jennifer Taylor, Jeff Tester, Jasna Tomic, Walter Vergara, Martin Vorum, Wang Tao, Liz Westcott, Yao Tandong, and Walter Youngquist.

As always, we are in debt to our editor, Linda Starke, who brings over 30 years of international experience in editing environmental books and reports to the table. She has brought her sure hand to the editing of not only this book but all my books during this period.

The book was produced in record time thanks to the conscientious efforts of Elizabeth Doherty, who prepared the page proofs under a very tight deadline. The index was ably prepared by Kate Mertes.

We are supported by a network of dedicated translators and publishers in some 23 languages, including all the major ones. In addition to English, our books appear in Arabic, Bulgarian, Chinese, Farsi, French, German, Hindi, Hungarian, Italian, Japanese, Korean, Marathi, Norwegian, Polish, Portuguese, Romanian, Russian, Slovenian, Spanish, Swedish, Thai, and Turkish. There are three publishers in English (United States/Canada, U.K./Commonwealth, and India/South Asia), two in Spanish (Spain and Latin America), and two in Chinese (mainland and Taiwan).

These translations are often the work of environmentally committed individuals. In Iran, the husband and wife team of Hamid Taravati and Farzaneh Bahar, both medical doctors, head an environmental NGO and translate EPI's publications into Farsi. Their translation of *Plan B* earned them a national book award. The ministries of environment and agriculture regularly purchase copies in bulk for distribution.

In China, Lin Zixin, who has arranged the publication of my books in Chinese for more than 20 years, enlisted a great team—the Shanghai Scientific & Technological Education Publishing House and WWF-China—for *Plan B 4.0*. Both Premier Wen Jiabao and Pan Yue, Deputy Minister of the State Environmental Protection Administration, have quoted *Plan B 2.0* in public addresses and articles. The Chinese edition of *Plan B* received a coveted national book award in 2005 from the National Library of China.

In Japan, Soki Oda, who started Worldwatch Japan some 20 years ago, leads our publication efforts and arranges excellent book promotional media interviews and public events. He is indefatigable in his efforts. The Kurosawa brothers, Toshishige and Masatsugu, distribute thousands of copies to Japanese opinion leaders.

Gianfranco Bologna, with whom I've had a delightful relationship for over 25 years, arranges for publication of our books in Italy. As head of WWF–Italy, he is uniquely positioned to assist in this effort. He is joined in the translation effort by a team headed by Dario Tamburrano of the Amici de Beppe Grillo di Roma.

In Romania, we are aided by former President Ion Iliescu, who started publishing our books 23 years ago when he headed the publishing house Editura Tehnica. Now he personally launches our books in Romania, ably aided by Roman Chirila, the current editor at Editura Tehnica.

In Turkey, TEMA, the leading environmental NGO, which works especially on reforesting the countryside, has for many years published my books. Inspired by Ted Turner, they distributed 4,250 copies of *Plan B 3.0* to officials, academics, and other decisionmakers.

In South Korea, Yul Choi, founder of the Korean Federation for Environmental Movement and now head of the Korea Green Foundation, has published my books and oversees their launching through Doyosae Books Co.

Most remarkable are the individuals who step forward out of seemingly nowhere to publish and promote our books. For instance, Lars and Doris Almström have now translated two editions in the *Plan B* series, and arranged for their publication in Swedish. They now actively are working to implement Plan B in Sweden.

Pierre-Yves Longaretti and Philippe Vieille in France literally accepted the call to action in *Plan B 2.0* and not only translated the book but engaged a world-class publisher, Calman-Lévy. They further established an NGO (Alternative Planetaire) and a Web site to promote Plan B for France (www.alternativeplanetaire.com).

Bernd Hamm, a professor at the University of Trier, arranged for a German publisher, Kai Homilius Verlag, to publish *Plan B 2.0*. Kai Homilius has now published *Plan B 3.0* and *Plan B 4.0*.

The Spanish editions of *Plan B 2.0*, *Plan B 3.0*, and *Plan B 4.0* in Latin America were spearheaded by Gilberto Rincon of the Centre of Studies for Sustainable Development in Colombia.

The Hungarian editions of Plan B 3.0 and Plan B 4.0, available electronically on our own Web site, are the result of the tireless efforts of David Biro, a school teacher in Hungary.

Those who are working to promote Plan B (see "People in Action" on our Web site) are gaining in both numbers and momentum.

We are also indebted to our funders. Without their support this book would not exist. Among these are the

Foundation for the Carolinas; the United Nations Population Fund; the Farview, Flora L. Thornton, Shenandoah, Summit, and Wallace Genetic foundations; and the craigslist Charitable Fund.

Earth Policy is also supported by individual donors. I would like in particular to thank for their major contributions Ray Anderson, Doug and Debra Baker, Tiziano Ciampetti, Junko Edahiro, Judith Gradwohl, Maureen Kuwano Hinkle, Betty Wold Johnson, Sarah Lang, Elena Marszalek, John Robbins, Peter Seidel, Emily Troemel, the Del Mar Global Trust, Jeremy Waletzky, and many others.

Finally, my thanks to the team at W. W. Norton & Company: Amy Cherry, our book manager; Devon Zahn, who put the book on a fast-track production schedule; Chin-Yee Lai, book jacket designer; Bill Rusin, Marketing Director; and Drake McFeely, President, with special thanks for his support. It is a delight to work with such a talented team and to have been published by W. W. Norton for more than 30 years.

And thanks to you, our readers. In the end, the success of this book depends on you and your help in implementing Plan B.

Lester R. Brown

About the Author

Lester R. Brown is President of Earth Policy Institute, a nonprofit, interdisciplinary research organization based in Washington, D.C., which he founded in May 2001. The purpose of the Earth Policy Institute is to provide a plan for sustaining civilization and a roadmap of how to get from here to there.

Brown has been described as "one of the world's most influential thinkers" by the *Washington Post*. The *Telegraph of Calcutta* called him "the guru of the environmental movement." In 1986, the Library of Congress requested his papers for its archives.

Some 30 years ago, Brown helped pioneer the concept of environmentally sustainable development, a concept embodied in Plan B. He was the Founder and President of the Worldwatch Institute during its first 26 years. During a career that started with tomato farming, Brown has authored or coauthored many books and been awarded 25 honorary degrees. With books in more than 40 languages, he is one of the world's most widely published authors.

Brown is a MacArthur Fellow and the recipient of countless prizes and awards, including the 1987 United Nations Environment Prize, the 1989 World Wide Fund for Nature Gold Medal, and Japan's 1994 Blue Planet Prize for his "exceptional contributions to solving global environmental problems." More recently he was awarded the Presidential Medal of Italy, the Borgström Prize by the Royal Swedish Academy of Agriculture and Forestry, and the Charles A. and Anne Morrow Lindbergh award. He has been appointed to three honorary professorships in China, including one at the Chinese Academy of Sciences. He lives in Washington, D.C.

If you have found this book useful and would like to share it with others, consider joining our

Plan B Team.

To do so, order five or more copies at our bulk discount rate at www.earth-policy.org

This book is not the final word. We will continue to unfold new issues and update the analysis in our

Plan B Updates.

Follow this progress by subscribing to our free, low-volume e-mail list or RSS feeds at www.earth-policy.org, and follow us on Twitter (@EarthPolicy) or on the Earth Policy Institute Facebook page.

Past Plan B Updates and all of the Earth Policy Institute's research, including this book with full endnotes and data sets, are posted at www.earth-policy.org for free downloading.

www.earth-policy.org